בס"ד

Israeli Vipers F-16 A/B Netz.
ISBN: 978-88-95011-20-2

Text, color profiles, graphic design and graphic layout: Amos Dor – Milano – www.iafe.net
English proofreading: Hans Wilms, Tamar Dor.
All the photos & graphics in this book are from Amos Dor private collection & art work, unless otherwise noted.

Publisher: **RN Publishing S.a.s. di R. Niccoli & C.**
Via Torelli 31 – 28100 Novara (NO) – Italy
Tel. & Fax. +39-0321-455108
www.rnpublishing.com

Print: Alcione S.r.l. – Lavis (TN) - Italy.

Front and back cover photos: Major Ofer

© Copyright 2021 – RN Publishing S.a.s. and Amos Dor.
All rights reserved. No part of this work covered by copyright hereon may be reproduced or used in any form or by any means - graphic, electronic, or mechanical, including photocopying, recording, taping or information storage & retrieval systems, without the written permission of RN Publishing & Amos Dor.

RN Publishing

TABLE OF CONTENTS

Dedication	4
Acknowledgments	5
Opening Remarks Major General Amir Eshel (res.)	6
The "Netz" period in my life Colonel Guy Shalev-Shelly	7
Introduction	11
The First Jet 117 Squadron	24
The Knights of the North 110 Squadron	36
Operation Opera	48
The Negev 253 Squadron	56
The Golden Eagle 140 Squadron	76
The Phoenix 144 Squadron	117
The Defenders of the South 116 Squadron	142
The Flying Dragon 115 Squadron	191
Flight Test Center Squadron	213
F-16A/B Technical Specifications	217
ACE Project	218
Color Profiles Drawings	220
117 Squadron Air Victories table	230
110 Squadron Air Victories table	231
253 Squadron Air Victories table	232
Closing of the "Netz" Array	233

DEDICATION

I would like to dedicate this book to my late father Haim Dor, who dedicated his life to the Israeli Air Force and to the security of the State of Israel, to my wife Rinati and my children Tamar, Ruben and Haim Eitan for their support and encouragement, and to a special person, the late Lt. Col. Adi Ribon, whose history of life is presented here.

Adi Ribon (Poretzky-Rubin), was born on December 5th 1929 in Kaunas, Lithuania. In 1941 the Nazis occupied Lithuania, *Adi* and his family were forced to move to the Kovno (Kaunas) Ghetto. After three years *Adi* and his stepfather were transported to the Dachau Concentration Camp (Germany) and barely survived an additional year of torture, forced labor and starvation, followed by the death march. They were finally liberated by the allied forces on the 30th of April 1945.

Adi volunteered to join the Jewish forces fighting for the independence of Israel ("Gachal"), he was trained by the "Haganah" in the French Alps and sailed to Palestine under an alias to join Division 9 (Div' "Oded") fighting both on the northern and southern fronts. When the war was over, *Adi* joined the Air Force and played an important part in building the technical array of the IAF and trained generations of air force technicians for planes and supporting systems maintenance. *Adi* studied aircraft mechanics and served in the 101 and 105 squadrons until moving on to the Technical Air Force Academy – the "Techni", right after the Sinai war.

Adi served in the "Techni" for 25 years until he was appointed IAF's "Chief Instruction Officer" ("Kahadar"). During those years he participated in bringing the Dornier Do 27 ("Dror") planes from Germany, was the Deputy Air Force Attaché in the Israeli Embassy in Washington D.C. and played a part in the massive procurement done before the "Yom Kippur War". In his last position he headed the Israeli training program for the F-16A/B "Fighting Falcon" ("Netz") in the United States (1979-1982) and laid the technical foundation for the arrival of these planes in the IAF.

Adi served at the IDF for 35 years and retired as a Lieutenant-Colonel. He continued to work in the military industry until his retirement in 1995, afterwards he volunteered in the community, traveled all over Israel and gave lectures to thousands of high school students, soldiers in army bases and anyone who would lend an ear about the importance of the Jewish state to the Jewish people and the destiny of a nation without a state.

Adi Ribon has lived a full life, mostly as a free man in his own country, he passed away on November 22nd 2011. He and his wife *Chasida* have three children and 10 grandchildren keeping his legacy alive.

On August 18th, 2020, the Israeli Air Force flew over Dachau Concentration camp. Major T., Deputy Commander of Squadron 105, flying one of the two F-16D ("Barak") planes, carried in his pocket Dachau prisoner badge number 85581, a badge that belonged to *Adi's* father, *Mordechai Rubin*, without whom *Adi* would have not survived the horrible conditions, the slavery and hunger he experienced in the same camp 75 years earlier…

IAF's "Chief Instruction Officer" (1976-1979).

Hill AFB - First F-16 handed over to IAF (January 31, 1980).

General Dynamics Farewell Ceremony (July 1982).

ACKNOWLEDGMENTS

Special thanks to Major General Amir Eshel (res.), Brigadier General Giyora Even (Epstein) (ret.), Brigadier General Amir Nahumi (ret.), Brigadier General Shlomo Sas (ret.), Brigadier General Rafi Berkovich (Berko) (ret.), Brigadier General Elisha Hosman (res.), Colonel Oded Marom (ret.) from the Israel Air Force Association, Colonel Guy Shalev-Shelly (Res.), Colonel Eytan Stibbe (Res.), Colonel Ofer Safra (ret.), Lt. Col. Dubi Ofer (ret.), Lt. Col. Ofer Einav (res), Major E.G., Major Ofer, Major Dror Albuher - IAF magazine chief editor, Asher Rot, Yehuda Borovik, Lt. Col. Shlomo Levi (ret.), Colonel Amir Chodorov (ret.), Hans Wilms, Hezi Shmueli, Pini Elmakies, Alberto Mocchetti, Riccardo Niccoli, Giovani Colla, Sariel Stiller, Amit Agronov, Lior Kestner, Boaz Aharoni, Tsahi Ben-Ami, Moni Shafir, Ron Zeev, Maurits Even, Massimo Pieranunzi, Dolev Gottdiener, George C. Kuntz, Noam Ribon. The author wishes to thank also all the others who preferred to remain anonymous.

*T*he arrival of the F-16 in Israel significantly strengthened the Air Force and changed the face of the Middle East.

The 'Netz', Hebrew name of the aircraft, was a technological revolution and gave new dimensions to the concept of 'multi-role'. A nimble and beautiful plane.

Less than a year after the integration of the 'Netz', the Iraqi nuclear reactor was destroyed and Syrian fighter planes and helicopters were shot down over Lebanon. Through the 'Netz', Israel drew a red line related to proliferation of military nuclear capability in the Middle East. The Israeli air force regained air superiority after the 1973 war and has lifted it up to unprecedented level. Over the years, the 'Netz' aircraft participated in wars and operations in the various theaters, hundreds of pilots were trained to fly it and they fell in love with a fighter that fitted them like a glove. Over time, more advanced aircraft were integrated into the Air Force, like the F-15I Ra'am ('Thunder') and F-16I Sufa ('Storm'). Alongside those, the 'Netz' proved that the power in his soul was still fresh' relevant and lethal.

Thousands of ground crews, who professionally and devotedly maintained the world's first fly-by-wire aircraft, were full partners in the achievements and glorious chapters written by the 'Netz' fighters in the history of Israel and air warfare. The advanced characteristics that the 'Netz' contributed to the Air Force served as a solid base for the successful integration of advanced aircraft in the IAF, including the recent mighty F-35.

The concepts devised in the 'Netz' era, developed with the acquisition of innovative capabilities, most of which are of "blue and white" origin, broke boundaries and increased the strength of the Air Force to new heights. This special aircraft, developed in the light of the "Fighter Mafia" vision in the USA, has skyrocketed far beyond the imagination of its designers.

Major General *Amir Eshel (res.)*
IAF Commander 2012-2017

My "Netz" period.

Colonel Guy Shalev-Shelly (Res.) who was the first F-16 technical officer of the 110 Squadron and the head of the technical delegation sent to study the F-16 in the US tells about the "Netz" period in his life.

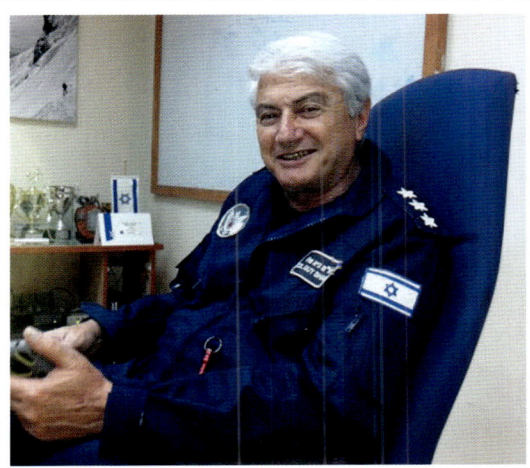

I served as the technical officer of the 110 "Ayit" (A-4 Skyhawk) Squadron for about eight months. It was a fascinating time. An old plane but a warhorse, which excelled in the Yom Kippur War. A squadron with a glorious tradition that in the past had operated Mosquito and Vautour aircraft. I performed the job with honors.

In January 1980 I went with my whole family to Hill AFB in the state of Utah, an incredibly beautiful country: snowy during winters and not too hot during summers. Its geographical location was ideal for traveling.

I was the commander of the delegation of technicians selected to study the F-16. The various courses we attended were not hard for us, because according to the American concept, one learns only what is necessary to know, therefore basic notions. The U.S. Air Force, in the event of unforeseen problems, had immediate access to the General Dynamics plant, the aircraft manufacturer, while ours did not have such an option. The solution was to learn and deepen the knowledge on our own. The guys would get great scores on the American tests but then they had to undergo the tests I made. The results were different, and the method was that whoever did not get a good score on my tests had to study on weekends instead of hanging out and traveling. The benefits of the method would be discovered later, when we returned to Israel.

The training period lasted about six months.

Prior to our return to Israel, we drove from Utah to Texas, to the aircraft manufacturer's factory, to complete the knowledge we lacked. At the end of the course we returned to Israel to build up the squadron. This included building and renovating all the infrastructure, writing training sets, placing the technicians and a lot more ... The joy of creating gripped us all. I was determined to establish an unparalleled squadron throughout the Air Force; different and with high standards in every field. Indeed, the return to Israel was in September 1980, and in February 1981 we were subjected to a very thorough audit of competence and alertness. The audit was unable to find even one flaw. At the end of the audit, Brigadier General *Yitzhak Geva*, head of the equipment division, found it appropriate to publish a letter throughout the Air Force saying that: "Such a level has never been observed in the Air Force ..." (I always keep the letter with me).

Two significant operational events characterized the period of my service in the 110 "Netz" Squadron.

1) Operation Opera (Attack on Iraqi nuclear reactor).

At the beginning of 1981, the squadron commander *Amir Nahumi* unexpectedly ordered me to prepare for deployment the following day with a pair of F-16s, for only one day to go to the Etzion AB, which is located to the south-west of Eilat. We were supposed to join another pair of the sister Squadron 117. Naturally and in an overall view of the activities in the squadron, I instructed my deputy *Zvika Barak* to choose a team for the descent with a Cessna plane to Etzion. For some reason, *Nahumi* insisted that I go down to deployment. *Nahumi* was a stubborn person so I did not try to oppose or make logical arguments. I went down with a team of four technicians and at the same time, a pair of F-16s (without armament) took off. We received them at one of the base's hardened aircraft shelters, refueled them, and checked them before the flight took off back to Ramat David AB, and we then followed them. Needless to say, I felt like a complete idiot.

In January, February, March and April, at the end of the training flights, the squadron performed many models of long range flights, sometimes with and sometimes without operational armament. I attributed this to what is called in the jargon of the aviation world "opening flight envelopes", which made perfect sense since it was a new aircraft. My suspicions arose - albeit with low intensity - when on some of the flights the IAF commander *David Ivri* joined in a two-seater plane and later even Chief of Staff *Rafael Eitan*. The squadron was busy receiving and training and *Nahumi* was constantly pressuring to urge reception of external fuel tanks from abroad. My curiosity increased but all my attempts to get a piece of information failed.

Another detail that seemed strange to me was related to the "foreign experts" who were at the base. To clarify, when the Americans sell an aircraft, the accompanying logistics package includes foreign experts who are supposed to assist in integration and to constitute a liaison with the manufacturing plant. Every flight would have to be reported to them. The report included aircraft number, configuration, flight duration and navigation system deviation (INS). On long-haul flights, I was asked to falsify the report so they did not suspect anything. On the eve of the second holiday of Pesach, early morning, a mission order was received: I had to prepare six F-16s in a configuration with three external fuel tanks and only two AIM-9L missiles for deployment that day to Etzion AB. I split the people into two groups: those who would prepare and launch the planes out of Ramat David AB and those who went down to Etzion to receive them. We selected the best planes by arranging a matrix, prepared an improvised logistics kit and flew with a C-130 to Etzion AB. At the same time the sister Squadron 117 performed the same operations. We received the planes, put them in one of the hardened aircraft shelters and only then we noticed that the base was empty: no one was there. We got the second mission order that included armament of each aircraft with two Mk.84 bombs, two AIM-9L type infrared missiles and maximum fuel (three external fuel tanks). Four aircrafts out of six were destined for takeoff. The AB security personnel fenced off the AB and we found ourselves in quarantine. No one could get in and no one could get out. We slept in the shelters. The delivery of the planes was at 15:00 the next day, which was a holiday. No word on the destination! The planes were supposed to perform "hot refueling" at the take-off position in order to get the maximum possible fuel. This operation is very dangerous and requires high management and skill.

Meanwhile I was thinking about the destination and trying to guess it. I suspected that the Air Force was about to attack the SAM missiles arrays that the Syrians had deployed in a rift in Lebanon, by flanking on the east, but it didn't make sense. The planes were prepared without any Electronic Warfare other than chaff/flare, and the F-4 squadron the Hammers had much more suitable means and armament for attacking ground missiles batteries.

At 16:00 the air crews arrived at the shelters, started the planes and then a cancellation order was received. Just before the fold, when curiosity was at its peak, I asked the AB commander, *Yiftach Spector*, to tell us the purpose of the deployment. We gathered in one of the shelter's domes and he told us that "The navy carried out an operation in Saudi Arabia in Bab Al-Mandab and we were on standby in case they got into trouble." Everything went well and we returned back to Ramat David.

Enough! I could not stand the curiosity anymore. I climbed in one of the F-16s' cockpit and looked up in the navigation system, all the path to the destination that the pilots did not bother to delete. I wrote them down on my cigarette pack and, in the squadron's operations room at Ramat David, I started following the flight path on a huge map on the wall. I was amazed! This led to Baghdad in Iraq! For the first time I learned where the "lunatics" were going to attack. For some reason I associated this with an article with pictures I read in the French Paris Match magazine which showed Iranian F-4 Phantoms attacking a nuclear reactor in Baghdad.

Things calmed down completely and the squadron returned to normal. In a small talk in the squadron commander's office, I expressed my dissatisfaction with the messy way we deployed, mainly regarding the logistical organization. Although this was the first deployment, it was contrary to the whole concept and standards we had worked hard in establishing in the squadron. No preparatory discussion, no thoughts, no search for excellence and departure from the norms of the Air Force, and no clear procedures. I asked if there was room for further reorganization and the answer was no. So I allowed myself to tell him "I had guessed" the place but not the target. He turned pale and made me swear to keep my mouth shut.

On the eve of Shavuot, that is to say about six weeks after the first deployment, while the squadron's air crews were in the process of preparing for a trip to the Sinai and the technical department was already home to celebrate the holiday, the scenario was repeated in a second edition. I avoided calling people to bring them back and preferred to pick them up from their home.

Our six pilots were:
Leader: *Lt. Col. Amir Nahumi. Squadron Commander*
No. 2: *Col. Yiftach Spector. AB Commander*
No. 3: *Major Relik Shapir. Deputy Squadron Commander.*
No. 4: *Captain Ilan Ramon. One of the planning team of the operation.*
No. 5: *Captain Shlomo Sas. First reserve.*
No. 6: *Captain Dubi Ofer. Second reserve.*

Some funny stories:
The chief of staff - *Raful* - arrived at the Etzion base to greet the pilots before their departure. He was still in the mourning period for his son who was killed with a Kfir plane from the same Etzion base. He brought each pilot a pack of dates just in case ... An amazing man.

A "Zamir" (Beechcraft Queen Air B-80) plane with intelligence and Mossad personnel landed but there was no one to receive it. The security guards allowed me to approach and accept it. One of the guys missed the step of the plane and fell down on the tarmac. The contents of the brown envelope he was holding scattered on the pavement and amongst other things there were bills of Iraqi dinars that were intended for pilots with the purpose of bribery ... I helped him collect the material not before taking from him one bill that has been with me ever since.

The AB commander, Col. *Yiftach Spector*, while strapped in the cockpit, asked me to exchange his ranks with me. I jokingly asked what to do with his ranks and he answered: "Save them, you may need them ...".

After being strapped in, *Nahumi* asked to come down in order to go to the bathroom. When he finished, out of superstition, he did not want to board the same plane. This of course complicated the whole order of exiting the shelter.

I strapped *Relik Shafir* in the cockpit myself. I decided

From Right to left: Ed Presley, Guy Shalev-Shelly, Lt. Col. Adi Ribon Training coordinator of the Israeli Air Force, Capt. Yaakov Nave and Bob Hayes at the ceremony of receiving certificates of completion of the F-16 course at Fort Worth Texas.
(Photo: Ron Zeev collection). ▶

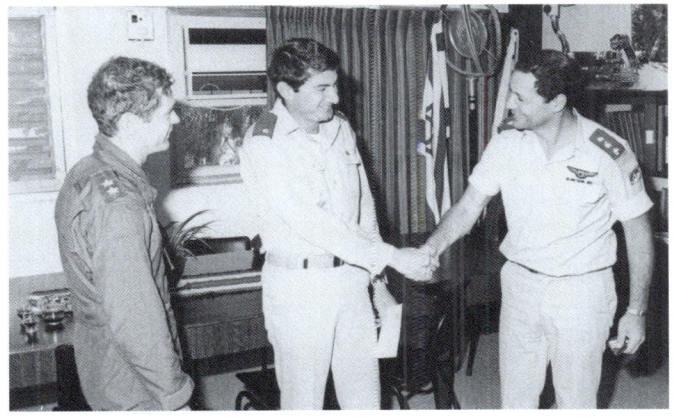

▲ *AB Commander Col. Yiftach Spector grants me the rank of Major in the presence of Squadron Commander Lt. Col. Amir Nahumi.*
(Photo: Guy Shalev-Shelly).

▲ *Technical fault handling. (Photo. Guy Shalev-Shelly).*

▲ *IAF ground crew group with Lt. Col. Adi Ribon Training coordinator of the Israeli Air Force at the ceremony of receiving certificates of completion of the F-16 course at Fort Worth Texas (Second from the left Zeev Ron).*
(Photo: Zeev Ron collection).

to give him a hint to make him understand that I knew where he was flying and in the flash of the moment I said to him: "Give them a greeting from the guys in Ramat Gan." He smiled and motioned for me with his finger on his lips to shut up.

At the hot refueling point at the end of the runway, *Relik* encountered a problem with the fuel system. The most important system together with the engine in this type of operation. In hot refueling, a refueling truck is connected to the plane and the plane "drinks" from the tanker and does not consume its internal fuel. For this purpose it is important that the pilot activates a certain switch in the cockpit. *Relik's* plane did not drink from the tanker. I first replaced the truck, but the problem persisted.

I connected myself to the internal communication system and asked *Relik* if he had activated the switch. He mumbled a vague answer. I started to get stressed. The whole thing happened in front of the first reserve pilot eyes, who called me and told me: "*Shelly*, is it true that *Relik* has a problem?" I answered "negative". And he tapped: "If you fix him a little glitch I'll give you everything you ask for!". Luckily we were given the order to align and take off. Still *Relik* came out with five hundred pounds less than the others. Crazy! When he returned after the raid I asked him why he had done that and with a smile typical of him he replied: "Do you think I would have given up because of such a mishap?"

The planes took off one after the other, attacked and destroyed the target and returned safely after about three nerve-wrecking hours that seemed like an eternity to me. The AB commander picked us up and explained to us that, without elaborating, we are partners in one of the Air Force's most daring and significant operations of all time.

Although we were told not to reveal anything on the mission, Kol Israel radio station announced that the Air Force had attacked the nuclear reactor that Saddam Hussein, the ruler of Iraq, had built in Baghdad with the help of the French government. It turned out that the reactor was supposed to go "hot" about three months later. The significance of the destruction of the reactor would gain its due importance about a decade later when the Americans attacked Iraq in the 1991 Gulf War.-

2) First world MiG kill with F-16.

In the same year, the Syrians deployed surface-to-air missile batteries in the Lebanese rift. The Air Force conducted a lot of reconnaissance sorties. For this purpose, the F-16 squadrons patrolled 24 hours to protect the photographers. In one of the launches, the "sister" squadron, 117 Squadron, managed to shoot down two Mi-8 Syrian helicopters in northern Lebanon.

One has to understand the atmosphere. The two squadrons were excellent (ours was better) and the competition between them was tough. It was clear to everyone that the squadron that valued better was the one who brought the "kills" first. And here they managed to shoot down two helicopters. The atmosphere in my squadron was so depressed, that my technicians did not dare to go to have their lunch since the shuttle bus would pass through the 117 Squadron and there they would have been mocked.

With each patrol I listened to the pilots' dialogue with control. In one of the sorties, a couple of our jets manned by *Amir Nahumi* and *Moshe Geffen* (a young pilot), constantly patrolled the valley until the fuel ran out. They were already on their way home when I suddenly heard on the radio - and only me in the entire operation room heard it - the magic sentence: "One knocked him down!". I jumped out of my seat. All the pilots around me who were present in the operations room and who did not listen to the radio, looked at me with compassionate looks, and were sure I was out of my mind.

I ran out and drove with my Vespa to the end of the runway to meet the planes that had just landed. And indeed *Nahumi's* plane was missing a missile! I signaled to the pilot: "Well ..?" And *Nahumi* ignores me. It took him an eternity to lift his thumb. That's it! The world's first MiG kill with an F-16!

The squadron was euphoric. The next day all the technicians went down to eat and as they passed by the sister squadron, they made a sign of a helicopter rotor with their finger, which meant: "you are chatting with helicopters but we shoot down MiGs".

In less than a year after having received the new jets, we had been able to establish a glorious squadron, carry out a historic attack on any scale, shoot down an enemy aircraft, and carry out further operational attacks. We got all the possible prizes of the IAF Force.

The first IAF F-16 patch (left), and the one designed by the Americans for the technicians (right).

Introduction

When General Dynamics was announced as the winner of the light-weight air superiority fighter F-X competition with their YF-16 design in the 1970's, the Israeli Air Force became immediately interested in what was to become the F-16 Fighting Falcon. The Israeli Air Force commanders were fascinated by the benefits of the new supersonic lightweight and economical fighter jet, which eventually evolved into an all-weather multi-role fighter. The first use of a relaxed static stability fly-by-wire flight control system is one of the characteristics which help to make the aircraft highly agile. This breakthrough in aircraft maneuverability was combined with a cockpit designed around the pilot from the inside out, another first, which excelled in improved and innovative human engineering, the bubble canopy that provides 360° all-round visibility, the ejection seat mounted at 30 degrees from the vertical to reduce the effect of the 9g acceleration forces acting on the pilot and the small side-mounted control stick mated with armrests that eases the control while maneuvering at high g as part of a fully-fledged HOTAS system.

The F-16 is armed with an internal M61 Vulcan cannon in the left-wing root and has 9 external stations (initially seven) for mounting a wide range of missiles, bombs, pods, and fuel tanks. A formidable Pratt & Whitney F100-PW-100 after-burning turbofan engine provided the aircraft with an immense thrust. The F-16 was the first fighter aircraft purpose-built to pull 9G maneuvers and can reach a maximum speed of plus Mach 2. The F-16's official name is "Fighting Falcon", but its pilots and crews usually call it "Viper" after the Colonial Viper in Battle-star Galactic and the perceived resemblance of the notorious snake.

The F-16 is the first production fighter aircraft intentionally designed to be aerodynamically unstable, also known as "relaxed static stability", which considerably improves maneuverability. Most aircraft are designed with positive static stability, which induces aircraft to return to a straight and level flight attitude if the pilot releases the controls; this reduces maneuverability as the inherent stability must be overcome. Aircraft with negative stability are designed to deviate from controlled flight and thus be more maneuverable, which is only possible with the use of a fly-by-wire system. At supersonic speeds, the F-16 gains stability (eventually positive) due to aerodynamic forces at those speeds.

Over 4,600 aircraft have been built since production was approved in 1976. Although the USAF no longer buys the F-16, improved versions are still being sold to countries around the world.

The then Israeli Air Force Commander, *Beni Peled*, announced "we want F-16s and a lot of them".

In August 1976, a first delegation led by IAF Commander *Beni Peled*, set out to the United States to test the new aircraft.

About a month later, it was decided that the F-16 would meet the requirements of the Air Force.

In March 1978, the incoming IAF Commander *David Ivri*, flew the F-16 for the first time and in August 1978, the first equipment contract was signed for 75 F-16s, 67 A models, and 8 B models. The Israeli name chosen for the F-16 was "Netz", which means "hawk" in English.

The first F-16s were supposed to arrive in Israel in the second half of 1981 after the supply of 160 aircraft ordered by Iran was completed.

But as we know, there are surprises in life and the revolution in Iran in January 1979 forced the Shah of Iran to escape to the United States and Ayatollah Khomeini, who had returned from exile, came to power. Subsequently, the relation between USA and Iran, which had become an Islamic state it is today, deteriorated and the Americans canceled the contract with Iran, offering the aircraft to Israel which was actually next in line.

This meant advancing the supply of aircraft to Israel by almost a year and a half.

The Israeli Air Force obviously jumped at the chance but demanded to make some changes to the planes according to their requirements, which the Americans initially did not agree to.

In May 1979, the first technicians destined for training departed for the United States and in November, a group of about 180 further technicians set out to study the aircraft. The training lasted about seven months and was held at General Dynamics plants.

The IAF decided to reactivate the "First Jet" (117) Squadron as the first Netz squadron with *Lt. Col. Zeev Raz* as squadron commander, *Dubi Yoffe* as first deputy squadron commander, *Hagay Katz* as second deputy squadron commander and *Relik Shapir* as systems officer. *Natan Snaron* was appointed as technical officer.

The "Knights of the North" (110) Squadron were selected to reactivate as the second Netz squadron, with *Lt. Col. Amir Nahumi* as squadron commander, *Udi Ben-Amitai* as first deputy squadron commander, *Relik Shapir* as second deputy squadron commander and *Shlomo Sas* as systems officer. *Gay Shalev (Shelly)* was appointed as technical officer.

The third squadron, "The Negev" (253) Squadron, was activated at a later stage as the third Netz Squadron and was to be based on the new Ramon AB in the Negev, which was under construction.

All three squadrons were initially stationed at Ramat David AB.

In February 1980, the IAF sent a first delegation of

four pilots on a two-month conversion course to Hill AFB Utah, that included 21 flights on the new aircraft. These pilots were: *Zeev Raz, Dubi Yoffe, Hagai Katz* and *Relik Shapir*. In March 1980, a second group was dispatched including: *Udi Ben-Amitai, Amos Yadlin, Dubi Ofer* and *Ilan Ramon*. In June 1980, a third group traveled to Hill AFB composed of: *Amir Nahumi, Rani Falk, Shlomo Sas* and *Arnon Sharabi*. *Shlomo Sas* recalls: *"During the conversion course in the USA we set out in a pair of F-16s to fight against a lone aircraft, an F-5 from their aggressor squadron.*

The F-5 "enters" on us, and a fight develops. Because the F-5 is inferior in performance, speed, and ability to turn, he is constantly pulling upwards in an attempt to break eye contact (the F-5 is a small aircraft, hard to see). During one of the F-5 climbs, a long trail of white smoke suddenly appears from the F-5 tail, and a few seconds later we saw a plane spinning .... We stopped breathing, waiting to see what happens.

Nothing, silence on the communication system... silence on the communication system... we noticed that the F-5 came out of the spin, the white smoke trail was gone, and we were waiting for the "knock it off" call, "stop the Dog Fight."... nothing... then we heard: 'resume fight' .... So we went ahead and tore it up .... this is how it is with the aggressors, as it turns out.

I do not remember that the incident was reported ...." The Israeli pilots immediately fell in love with the Netz. This is how *Zeev Raz* described it: *"A legendary plane, its human engineering is such that it does not need two people. It is a very energetic plane. She has a lot of energy in maneuvering, and that is her greatness. To this day the Netz is the best aircraft in air combat compared to any other aircraft."*

On July 2, 1980, about a year before the expected original delivery date in the IAF, the first four F-16s landed at Ramat David AB. Two F-16As with tail numbers 105 and 107, and two F-16Bs with tail numbers 008 and 015. The planes were flown on a direct flight from the US by American pilots. The flight lasted approximately about 11 hours, including multiple in-flight-refueling, due to political difficulties in obtaining a permit from Portugal for a stopover in the Azores islands. On subsequent transfer flights this obstacle no longer existed. The delivery operations were given the code name of "Peace Marble I". The last quartet of the 75 aircraft ordered landed in Israel at the end of October 1981.

As part of the yearly US military assistance to Israel, in 1994 the US administration agreed to supply Israel with 50 F-16s (36 model A, and 14 model B) from the U.S. Air Force surplus. These planes, named "Netz 2" by the IAF, arrived in Israel at the end of 1994 ("Peace Marble IV") and saw the "Phoenix" Squadron being reactivated.

During the 35 years of operation of the "Netz" in the IAF, the aircraft has undergone many improvements and upgrades. The planes underwent structural changes under the "Falcon F" program in order to extend the life of the aircraft thanks to various mechanical interventions, like replacing certain panels, and installing improved avionics systems such as a navigation system based on a laser gyroscope RLG (Ring Laser Gyro). The brake systems have been improved and anti-skid controllers have been installed in all aircraft. The F-16 A/B was the first in the Air Force to operate the improved Sidewinder AIM-9M missile. By the middle of 2006, the circle was closed, and all F-16 A/B aircraft in the IAF had been upgraded to the same specifications.

◀ *Captain Shlomo Sas is getting ready for his solo flight in a two-seater F-16. A flight that ultimately did not take place due to a flight control BIT malfunction. (Photo: Shlomo Sas collection).*

The first F-16B "Netz" Block 5 s/n 78-0355, IAF #001 during test flights in USA.

▲ *IAF ground crew group with Lt. Col. Adi Ribon, training coordinator of the Israeli Air Force. (Photo: Zeev Ron collection).*

"Peace Marble 1" F-16A

IDF/AF S.N.	C.N.	BLOCK	USAF S.N.		117 Sqn. Name
100	6V-01	5	78-0308	21.9.1980	Carish
102	6V-02	5	78-0309	17.9.1980	Meteor
105	6V-03	5	78-0310	2.7.1980	Sa'ar
107	6V-04	5	78-0311	2.7.1980	Sufa
109	6V-05	5	78-0312	10.8.1980	Nesher
111	6V-06	5	78-0313	10.8.1980	Yahalom
112	6V-07	5	78-0314	17.9.1980	Ra'am
113	6V-08	5	78-0315	17.9.1980	Herev
114	6V-09	5	78-0316	21.9.1980	Snapir
116	6V-10	5	78-0317	9.11.1980	Shavit
118	6V-11	5	78-0318	20.10.1980	Tsiltsal
121	6V-12	5	78-0319	20.10.1980	Barak
124	6V-13	5	78-0320	24.11.1980	Kochav
126	6V-14	5	78-0321	24.11.1980	Magen
129	6V-15	5	78-0322	24.11.1980	Hanit
131	6V-16	5	78-0323	24.11.1980	Lahav
135	6V-17	5	78-0324	11.12.1980	Tzor
138	6V-18	5	78-0325	11.12.1980	Shahak
219	6V-19	10	78-0326	11.12.1980	
220	6V-20	10	78-0327	11.12.1980	
222	6V-21	10	78-0328	11.12.1980	
223	6V-22	10	78-0329	18.2.1081	
225	6V-23	10	78-0330	21/01/81	
227	6V-24	10	78-0331	21/01/81	
228	6V-25	10	78-0332	21/01/81	
230	6V-26	10	78-0333	21/01/81	
232	6V-27	10	78-0334	21/01/81	
233	6V-28	10	78-0335	25/02/81	
234	6V-29	10A	78-0336	25/02/81	
236 - (140)	6V-30	10A	78-0337	18/03/81	
237	6V-31	10A	78-0338	18/03/81	

IDF/AF S.N.	C.N.	BLOCK	USAF S.N.		117 Sqn. Name
239	6V-32	10A	78-0339	18/03/81	
240	6V-33	10A	78-0340	30/03/81	
242	6V-34	10A	78-0341	30/03/81	
243	6V-35	10A	78-0342	18/03/81	
246	6V-36	10A	78-0343	23/04/81	
248	6V-37	10A	78-0344	23/04/81	
249	6V-38	10B	78-0345	24/04/81	
250	6V-39	10B	78-0346	24/04/81	
252	6V-40	10B	78-0347	27/05/81	Keshet
254	6V-41	10B	78-0348	20/05/81	Noga
255	6V-42	10B	78-0349	27/05/81	
257	6V-43	10C	78-0350	27/05/81	Kidon
258	6V-44	10C	78-0351	20/05/81	Romach
260	6V-45	10C	78-0352	27/05/81	
261	6V-46	10C	78-0353	28/08/81	
264	6V-47	10C	78-0354	31/08/81	
265	6V-48	10C	80-0649	31/08/81	
266	6V-49	10C	80-0650	31/08/81	
267	6V-50	10C	80-0651	17/07/81	
269	6V-51	10C	80-0652	17/07/81	
272	6V-52	10C	80-0653	17/07/81	
273	6V-53	10C	80-0654	17/07/81	
274	6V-54	10C	80-0655	17/07/81	
275 - (115)	6V-55	10C	80-0656	17/07/81	#115 from 18.12.2012
276	6V-56	10C	80-0657	28/08/81	
277	6V-57	10C	80-0658	31/08/81	
281	6V-58	10C	80-0659	28/08/81	
282	6V-59	10D	80-0660	28/08/81	
284	6V-60	10D	80-0661	22/09/81	
285	6V-61	10D	80-0662	18/09/81	
287	6V-62	10D	80-0663	18/09/81	
290	6V-63	10D	80-0664	18/09/81	
292	6V-64	10D	80-0665	23/10/81	
296	6V-65	10D	80-0666	23/10/81	
298	6V-66	10D	80-0667	23/10/81	
299	6V-67	10D	80-0668	23/10/81	

"Peace Marble 1" F-16B

001	6W-01	5	78-0355	20.10.1980	
003	6W-02	5	78-0356	31.01.1980	Iftach
004	6W-03	5	78-0357	28.2.1980	
006	6W-04	5	78-0358	9.11.1980	
008	6W-05	5	78-0359	2.7.1980	Hetz
010	6W-06	5	78-0360	10.8.1980	
015	6W-07	5	78-0361	2.7.1980	
017	6W-08	5	78-0362	10.8.1980	

F-16A "Netz" Block 5 s/n 78-0309, IAF #102 during a test flight in USA and in advertisement photo with tail #117.

◄ IAF Ground crew during the training period at Hill AFB. (Photo: Tommy Hammer collection).

▲ The first four F-16s to leave the production line were transferred to the Hill AFB for the purpose of conducting conversion courses for the first IAF pilots.

F-16B #003 & 004 during test flights at the General Dynamics plant. Please note the SN 78-0356 which is clearly visible & the USAF roundels still on the jet. ▶
(Photo: Major E.G. collection).

◄ January 31, 1980 - The first aircraft delivery ceremony flown from General Dynamics plant in Fort Worth Texas to the U.S. Air Force Hill AFB in Utah. From right to left: Mr. Richard E. Adams, CEO of General Dynamics, Aryeh Levy, head of the Israeli Department of Defense delegation, General James Alan Abrahamson, director of the F-16 Multinational Air Combat Fighter Program and Brigadier General Amos Lapidot, who was then head of the IAF air division and flew the F-16 with an American pilot in the back seat.

◄ Brigadier General Amos Lapidot, talking with Mr. Richard E. Adams, CEO of General Dynamics, and General James Alan Abrahamson, director of the F-16 Multinational Air Combat Fighter Program. A few minutes before take-off to Hill AFB.

Brigadier General Amos Lapidot, Immediately after landing at Hill AFB in Utah.
(Photos: Zeev Ron collection). ▼

January 31, 1980 - Brigadier General Amos Lapidot, descending from the cockpit for the delivery ceremony. ▶

Amos Lapidot Receiving the "F-16 Flight Certificate" from a representative of General Dynamics. In the back: David Schlesinger, the armament officer of 117 Squadron, on the left: Haim Levy the management officer of the IAF delegation. ▶

Amos Lapidot talks with the delegation of Israeli Air Force ground crews who arrived to welcome him and the first F-16. (Photos: Zeev Ron collection). ▶

F-16A #107 and #105 and F-16B #008, #010 and #015 (Block 5) during test flights, including in-flight refueling, at the General Dynamics plant in Fort Worth Texas (formerly Carswell Air Force Base). (Pages 20-21). (Photos: Zeev Ron collection).

21

*F-16A #2XX (Block 10) during test flights, at General Dynamics plant in Fort Worth Texas (formerly Carswell Air Force Base) Note on the left, F-16/79 (No. 50752) intermediate fighter prototype. (Pages 22-23).
(Photos: Zeev Ron collection).*

Note the manufacturer serial number painted on the air intake.

The First Jet Squadron

On June 17, 1953, the first two jet aircraft landed at Ramat David AB. These were a pair of new British Gloster Meteors that opened the jet era in the IAF, and the new 117 First Jet squadron was established specifically for them.

On February 1, 1962, the 117 Meteor Squadron was officially decommissioned to be reactivated again on July 7 of that same year with the new Dassault Mirage III aircraft.

On October 12, 1979, after 17 years of service and with 94.5 kill credits, the 117 squadron again closed, but not for long. On Wednesday, July 2, 1980, the squadron was reactivated anew, this time as the first F-16 Fighting Falcon squadron in the IAF.

The arrival of the F-16s was accompanied by a modest welcoming ceremony, where squadron commander *Lt. Col. Zeev Raz* reviewed the squadron's glorious history.

Squadron commander *Lt. Col. Zeev Raz* decided to give names to all the squadron's F-16s, like Israeli prime minister David Ben-Gurion named the squadron's Meteor Jets in 1953. Netz #105 was named Sa'ar and #107 was named Sufa, exactly as David Ben-Gurion named the first two Meteor Jets.

The F-16 maiden flight in the skies over Israel was performed by *Lt. Col. Zeev Raz* on July 8, 1980 in Netz #105. Immediately afterwards, the squadron commenced regular training flights, initially with the pilots of the first and second conversion course, later the pilots of the third conversion course joined them.

The squadron was tasked with three main objectives: training a large number of pilots to staff three Netz Squadrons, achieving operational status as quickly as possible in view of the security tensions with Syria over the Lebanese issue, and the third and secret mission, preparing the F-16s to attack the Iraqi nuclear reactor.

The squadron continued to receive new F-16s at a rate of 2-4 each month, and as early as November the squadron was declared operational.

In September 1980, six pilots left the squadron and moved to establish the sister squadron - 110 squadron: squadron commander *Lt. Col. Amir Nahumi*, his deputies *Ehud Ben-Amitai* and *Relik Shapir* and the pilots *Shlomo Sas, Arnon Sharabi* and *Dubi Ofer*. At that time, the first conversion course in

The first four F-16s after landing at Ramat David AB on Wednesday, July 2, 1980.

▲ Preparing the new jets for the ceremony.

The IAF commander Major General David Ivri speaks at the reception ceremony of the F-16s. ▶
(Photo: Zeev Ron collection).

Ground crews stand next to the new planes at their reception ceremony. ▶
(Photo: Ron Zeev collection).

Israel was started, in which *Zvi Vered, Amos Mohar, Yishai Spector, Ofer Safra, Mordechai Rader, Eliezer Shkedi, Gideon Dor, Yizhak (Sasha) Levin* and *Rafi Berkovich (Berko)* took part.

First attacks and air combats - Following an attempt by Christian forces in Lebanon to occupy the city of Zahla, which was in the heart of the Syrian occupation and influence zone, leading to a direct clash between the Christian Phalanges and the Syrian army in April 1981, the Christians turned to Israel for assistance. On April 27, 1981, the squadron launched their first attack of the Netz era. *Zeev Raz, Gideon Dor, Hagai Katz* and *Yishai Spector* attacked a bridge in central Lebanon to block Syrian army reinforcements. Each of them dropped six Mk.82 bombs.

The first aerial combat victory was soon to be a fact.

April 28, 1981 - *Rafi Berkovich (Berko)*, a young pilot in the squadron, shot down a Syrian Mi-8 helicopter (F-16A #112). *Berko* recalls: *"At the time I was on a regular training day, after already having been training for five months flying the F-16. At that time there were a lot of scrambles against Syrian MiGs. Both sides approached the border, but we did not engage each other. The scramble that day was following the entry of Syrian helicopters that provided supplies to their forces in Lebanon against the Christians. Israel had decided to intercept the line of helicopter supplies. That day I was scheduled for standby for a 5-minute takeoff for intercept missions, squadron commander Lt. Col. Zeev Raz was the leader, and we were kicked right out from the coffee break in the squadron club. We flew at low level along the coast to Beirut, overflying way-points along the way, the tension was high amongst us all. Over Beirut we climbed to 20,000 feet. Control said there were helicopters in the Riak area northeast of Beirut, and we had to try to intercept them. We set up for the first intercept. We had to get into the surface-to-air missiles defended area, so we asked for permission from the controller. The controller did not give approval. We bypassed the surface-to-air missiles defended area and the radar lock was lost. We turned back, went in again, and this time the controller gave permission to enter the surface-to-air missiles defended area air space, and once the helicopter was identified with certainty as a Mi-8, we got permission to fire. I had IR lock issues and was unable to lock despite all attempts. I decided to launch a missile anyway, the missile hit the ground and blew up a small shack in what was an agricultural area, By the way it was the first missile I ever launched in my life.*

We started again turning in at high altitude and the controller announced that there are MiGs twenty miles

▲ Capt. Ofer Safra soaking, after the traditional wash-down on his return from his first solo flight on F-16.

◄ Participants in the first conversion course held in Israel. From right to left: Yishai Spector, Rafi Berkovich (Berko), Zvi Vered, Eliezer Shkedi, Gideon Dor, Mordechai Rader, Ofer Safra, Amos Mohar and Yizhak (Sasha) Levin. (Photos: Ofer Safra collection).

from us, heading in from the East and within a two-minute reach. I switched to guns, flying at exceedingly high speed in a diving attack, fired a burst of half the amount of the 250 shells. The helicopter caught fire at the end of the burst and crashed into the ground. The whole base waited beside the runway to greet the first squadron's and worldwide F-16's intercept success."

April 21, 1982 - A pair of Syrian MiG-23s approached the Lebanese coast north of Beirut. In doing so, the Syrians deviated from the agreement between Israel and Syria, according to which the undeclared border between the two countries coincides with the border of the Syrian surface-to-air missile protected air space in the Beqaa Valley.

Two pairs of F-16s that were on standby scrambled to intercept the invaders. The leader of the first formation, *Hagai Katz* (F-16A #284), identified them as MiG-23s. *Hagai* easily got on the MiG's tail, launched an AIM-9L Sidewinder which flew straight into the target and the MiG exploded. This was the first MiG-23 kill in world history. The second MiG escaped east into the surface-to-air missile protected air space. *Zeev Raz*, the squadron commander, with his wingman *Yohai Eshkol*, chased the MiG. *Yohai* launched an AIM-9L Sidewinder but the missile missed. *Raz* (F-16A #107) launched an AIM-9L Sidewinder and immediately broke to return to Israeli territory without knowing the fate of the MiG. Only afterwards *Yohai* told him that he saw the MiG going down.

The squadron did not have to wait long for further kills. In the afternoon, another pair was scrambled (*Maj. Dubi Yoffe* and *Lt. Sasha Levin*), and *Dubi Yoffe* in F-16A #126 downed another Mi-8 over Jabel-Snin, this time by launching an AIM-9L Sidewinder. This was the beginning of the snowball that culminated in the First Lebanon War (Operation Peace for Galilee)".

June 6, 1982 Operation Peace for Galilee - Following the terrorist assassination attempt of the Israeli ambassador to Britain, who was seriously wounded and remained paralyzed, the Israeli Air Force attacked terrorist targets in Lebanon, and the Israeli government decided to launch a military operation with the aim of occupying southern Lebanon and destroying terrorist forces, with the scope to restore peace for the northern towns and cities in Israel. At the time, 117 squadron was the spearhead of the Israeli Air Force. The squadron had 12 operationally qualified pilots, many aircraft, and was based at Ramat David AB, which is the closest base to the border with Lebanon. The pilots sat for hours in the ready room on standby or inside the jets with running engines, ready for scramble.

During the operation, only two sorties from the squadron were assault missions, all other sorties were air-to-air missions. *Berko* recalls: *"In the first days of the operation, the squadron did not shoot down even one aircraft. We scrambled a lot, but until the fourth day there were no kills"*.

June 9, 1982 - The fourth day of the fighting, a four ship F-16 formation from the squadron (code name: Tchehi) took off with *Zeev Raz* in the lead. Number 2 was *Eytan Stibbe* (F-16A #129) number 3 *Shkedi* (F-16A #107) and number 4 *Danny Oshrat*. Stibbe recalls: *"We patrolled out over the sea north of Beirut when the controller directed us east for an intercept in the heart of the surface-to-air missile defended zone. It was the day the Israeli Air Force had destroyed the missile batteries, but we still did not know about that. We identified a pair of MiG-23s that apparently, judging by their flight paths, also did not know that*

▼ Lieutenant Rafi Berkovich carried on the shoulders of his friends upon his return from the world's first F-16 kill".

▼ F-16A #112 with which Rafi Berkovich shot down the Syrian Mi-8 is seen here with two kill marks on it.

the batteries were destroyed. *"I passed at a very low speed"*, continues *Stibbe*, *"and you think, what do I do now? What are they up to? What are they thinking? Due to the F-16's performance the encounter was truly short. I aimed and punched off a missile. It was the first life missile launch in my career, and the time that elapsed from the moment the pickle was pressed until the missile went off the rail seemed eternal to me. The MiG was hit, the Syrian pilot ejected, and passing by him while he was hanging on the parachute was a tremendous experience. The second MiG was shot down by Shkedi".*

After landing and processing the debriefing data, it turned out that *Shkedi* and *Stibbe* hit the first MiG-23 together and each was credited with half a kill.

June 10, 1982 - The squadron continued to carry out CAP (Combat Air Patrol) missions to protect IDF forces and the CAS (Close Air Support) attack aircraft that supported them. A four ship F-16 formation scrambled and waited in a standby holding pattern. *Rafi Berkovich (Berko)* remembers: *"Zvika Vered led the formation, I was nr. 2 (F-16A #116), Eliezer Shkedi nr. 3 and Sasha nr. 4 (F-16A #111). We patrolled over Lake Qaraoun. We kept getting alerts from control that MiGs were incoming. At one point we formed an "Indian circle" at 16,000 feet patrol altitude, when control told us to look for low altitude 'cockroaches' (nickname for MiG-23s). I looked down and spotted a pair of MiGs. I was the only one in the formation to see them. I asked the leader to 'cut' in front of him and requested to open fire, because I identified them at long range, about 1.5 mile. I got clearance to launch and immediately fired at both MiGs. Launched the first missile and tracked it - I had already launched missiles before, but it remains hypnotizing – focusing my eyes, the distance was such that after 5 seconds I thought the missile had missed. So, I turned back into them and then came across his number 2 and launched on him. I downed the two MiGs but at that moment I only saw the first one go up in a fireball. While the second missile was still flying, I got into a 'sandwich' between a pair of MiG-21s. They flew behind me in tandem, but I did not see them. Sasha, flying about two miles behind me, did not recognize if I was number 1 or 2 and said: 'On the right side of someone there is a MiG'. I looked and saw a full-size MiG-21, about 200 meters next to me at my three o'clock. The speed was tremendous, 600-650 knots. I did two scissor maneuvers with him at a very high g-load (8.6g), I could barely hold my edge, but in the end, I succeeded in getting the angle on him. I fired two bursts, the first one missed, but the second burst hit home ... The MiG went up in a fireball right front of me. At this point I thought I downed two MiGs. I did not see the second MiG explode, because I was already in the dogfight with that third MiG. Meanwhile, the MiG-21 that flew with the MiG I downed was after Shkedi. Shkedi spotted him and reported that there were MiGs behind us, we all broke away like crazy, and then Sasha knocked him down. In fact, we were four up against four, and we downed 4 MiGs. On the way back, after we burned some fuel, we continued to patrol some more. Sasha and Shkedi had a little more fuel left, and Sasha shot down another Mi-8 helicopter with a missile. In fact, the results of the combat came only in the debriefing after two days. When we went through the films, we saw that there were more fires than reported kills. Eytan Stibbe told me 'Look, maybe you got another kill'. We asked intelligence and they reported the whole quartet had been shot down. So, I got the third kill too. As far as I know this is a Guinness world record. Shooting down 3 aircraft in 45 seconds."*

Later that the day, *Shlomo Zeitman* (F-16A #124), *Ami Lustig* (F-16A #138) and *Hagai Katz* (F-16A #118) shot down one MiG-23 each.

June 11, 1982 - Officially, it was the last day of the operation. During the morning, several air battles took place and in one sortie a formation of four F-16s from the squadron with call sign "Mavreg" (screwdriver) set a record of downing 9 enemy aircraft.

Eytan Stibbe remembers: *"The leader of the formation was Rani Falk (F-16A #258), with Dan Oshrat as nr. 2 (F-16A #254), Amos Bar in nr. 3 (F-16A #252) and me as nr. 4 (F-16A #107). The mission was to patrol over Raik. The first aircraft sighted was a MiG-21, which survived a battle in the southern Beqaa valley with a 110 Squadron formation and fled north. We turned in on him, and Falk downed him with cannon fire. We continued our patrol and then encountered MiG-23s and Su-22s heading in from the east to attack our ground forces. At this point we split. I caught a MiG-23 from the south. I got the angle on him, launched within 2 miles, and tracked the missile path. Before it hit, another MiG-23 passed under me at remarkably close range and low altitude. I managed to see the first MiG crashing, and went immediately after the second MiG. When I was 600 meters behind it, I hit the pickle and launched from 500 meters. The noise of the missile was very loud. It flew straight in and hit the MiG, which crashed into the ground. I pulled up hard from a low altitude to not pass through the fireball.*

At this point I discovered that I was alone, and my formation was looking for me to the north. I started climbing from a low altitude, defending myself from ground missiles, when I discovered another target on my radar. I looked up and saw a pair of fighter-jets a mile and a half in front of me. I reported them with control and set out after them. At first, I could not identify the type of the jets, and only when I passed over them, I recognized them as Su-22 fighter-jets. I went after them east-bound, almost up to the Syrian border. Instructions were not to engage aircraft in Syrian airspace. Luckily, their number 2 turned back in a spiral dive. I turned on the HUD camera, and went for a low-altitude cannon attack. I fired a burst at a

The first two F-16s in the world to be credited with MiG-23 kill.

F-16A #284 that until its retirement from active service in the IAF was credited with a single kill ▶

F-16A #107 the world leading F-16 MiG killer with 6.5 credits. ▶
(Photos: Moni Shafir)

F-16A #138 named "Shahak" was credit with a single kill (10.6.1982, Ami Lustig - MiG-23). ▶

F-16A #254 named "Noga" was credit with two kills. Seen here before taxing for take-off. ▶

range of 150 meters and the Su-22 was hit. I pulled up from the dive, and luckily, I got in the clear at very low altitude. At this point I had shot down 3 aircraft in about two minutes.

But the battle was not over yet. I found my formation, joined up, continued to patrol over Raik, and got an automatic radar lock on a slow target. We circled over it several times but failed to identify the target. Then Dan Oshrat, the squadron champion in aircraft identification, announced that it was a Syrian Gazelle helicopter. Falk climbed up to cover us, and the three young pilots attacked one after the other. I went for a missile attack with the AIM-9P. I got off the volley, flipped over to see if the missile hit, and reported: 'four launched, four hits'. The helicopter exploded into a fireball."

In this raid, *Rani Falk* downed one more Su-22, *Amos Bar* one Su-22 and *Dan Oshrat* one MiG-21 and one Su-22.

A world record off nine kills for a four-aircraft formation in one sortie.

July 13, 1982 - While flying east from Ramat David AB at 1,500 feet altitude, *Shlomo Zeitman* in "Netz" #252 lost engine power. *Shlomo* dropped the external centerline fuel tank and managed to bring the aircraft in for a perfect landing at Megiddo Airport. For this Shlomo received a citation.

October 10, 1982 - Official opening of the "Netz" Flight Simulator Division. As part of the F-16 fighter-jets deal, the Air Force purchased a flight simulator for the aircraft. In April 1982, the parts of the simulator arrived at Ramat David AB, and their assembly was completed inside a dedicated structure on October 10, 1982. 117 Squadron operated the "Netz" Flight Simulator for 6 years. On January 1988, the Flight Simulator was transferred to Ramon AB.

October 13, 1982 - Lieutenant Colonel *Zeev Raz*, the first squadron commander of the "Netz" era, hands over command of the squadron to Lieutenant Colonel *Itzik Gat*.

June 26, 1984 - Loss of F-16B #008. Lt. Col. *Sasha Levin,* who participated in the first F-16 conversion course that took place in Israel, tells about the incident in the Israeli Air Force magazine: *"I did a low-altitude roll, a relatively simple exercise, but then, all of a sudden, I lost 5,000 feet. I found myself 7,000 feet AGL, in a situation called 'deep stall'. The center of gravity of the F-16 aircraft is farther aft compared to other fighter jets. It was designed this way to improve its maneuverability, but this feature causes instability. The jet sank quickly. I knew that according to the procedures, I would have to eject at a minimum altitude of 6,000 feet. It took me a second to recover, and then I realized: 'Either I act, or it is not going to end well'. I did not have time to panic or pray. I began to extract from memory things that would help me. Then I remembered a piece of advice I had heard a few years earlier in a lecture by a test pilot at General Dynamics. 'Going into afterburner in such a situation will stop the jet from sinking'. Ignition of the burner is not self-evident. The angle of attack at which the jet was does not allow for a sufficient air flow to the engine and without it no AB combustion is possible. However, against all odds, the burner ignited. The sinking did not stop but was significantly retarded. I still knew there would not be much airspace left with the jet sinking so fast. I started rocking the aircraft, rocking it from side to side. This is a method I had read about in books but never practiced. Time was running out. I saw 2,500 feet on the altimeter and decided this was it. Need to put the manual down and get out. I pulled the ejection handle, and a light breeze blew in my face. The horizontal speed was almost zero - not much more than a car driving in city streets. I remember landing in the water a few seconds later, climbing in the dingy, and sunbathing for about half an hour until the SAR helicopter arrived.*

I landed in front of Hadera, in the distance I could make out the chimneys. When I got into the Super Frelon helicopter which came to rescue me, I was informed that the AB commander, Maj. Gen. (Res.) Herzel Bodinger, wanted to talk to me. Contrary to procedures, I was flown in wet and frozen for interrogation."

August 22, 1984 - Lieutenant Colonel *Itzik Gat* hands over command of the squadron to Lieutenant Colonel *Dubi Yoffe*.

June 13, 1985 - A Syrian DR-3 drone was identified over the Galilee. Lieutenant Colonel *Itzik Gat*, now attached to headquarters, was on a scheduled flight that day, performing a test flight in "Netz" #112. Towards the end of the flight, he was asked by control if he had enough fuel. His answer was yes, and was vectored to the drone. He spotted the drone on radar over the shoreline, got behind it at a speed of 480 knots and an altitude of about 3,000 feet. From approximately 200 meters he fired using the on-board cannon. The drone broke up in fragments and splashed into the sea.

October 8, 1986 - An F-16 formation from the First Jet Squadron was performing a dog-fight combat training. As part of the training, the pilots maneuver their jets aggressively to get on the tail of the aircraft they are chasing, and the pursued aircraft also maneuver sharply to avoid interception. At the beginning of one of these high-g turns, while performing an overflight to attack a SAM missile battery, Captain *Gal Weisberg* in F-16B #003 blacked out. The jet continued to turn sharply, went into a dive, and crashed.

During the final months of the "Netz" period, the squadron participated in a series of attacks on terrorist targets in Lebanon, while at the same time continuing its usual training activities.

November 27, 1986 - Closing ceremony the squadron's "Netz" era, with the flight of the last 12 F-16s of the squadron, which transferred to 140 Squadron at Ramon AB.

PHOTO GALLERY

▲ F-16A #102 at the Etzion AB during the preparations for the attack on the nuclear reactor in Iraq. (Photo: Major E.G. collection).

F-16A #105 shortly after its arrival, the squadron badge was already painted on it, but still without the "Sa'ar" (storm) name given to it. ▶

F-16A #107 name "Sufa" (storm) seen here during engine replacement. (Note please the fake Squadron emblem. Affixed to IAF planes when foreign visitors and photographers visited and the Air Force knew they might be published). ▶

◄ F-16A #111 after landing at its new home base - Ramat David AB - August 10, 1980.

▲ F-16A #112 during low altitude flight over Northern Israel. ▼

F-16A #118 shortly after its arrival, the squadron badge was already painted on it, but still without the "Tsiltsal" (Harpoon) name given to it. Landing at its home base - Ramat David ▶

F-16A #121 Shortly after its arrival, the squadron badge was already painted on it, but still without the "Barak" (Lightning) name given to it. ▶

F-16A #252 named "Keshet" (Bow) and F-16A #131 named "Lahav" (Blade) escorting F-4E Kurnass #225 from the 69 Hammers Squadron. ▶

▲ F-16A line up taken in mid 1980s.

▲ F-16B #003 named "Iftach" taxiing to the take-off position for another routine training flight.
(Photo: Major E. G. Collection).

▲ F-16B #004 during in-flight refueling from an IAI modified Boeing 707 tanker over the Mediterranean sea.

▼ F-16B #015 in flight over Israel, still with its original factory tail number.

The Knights of the North Squadron

In July 1953, the special element 110 was established within 109 Squadron for night fighter duties and training flights. This was the beginning of the 110 Squadron which on August 2, 1953 was promoted to squadron status within the Israeli Air Force. The squadron operated the Mosquitos until April 1, 1957 and contributed greatly to the outcome of the battle in "Operation Kadesh" in 1956.

On January 19, 1958, the squadron was reactivated, this time as the only Sud Aviation Vautour squadron in the Israeli Air Force. During the Six Day War, the squadron was credited with the first and only kill in the world with a Vautour. The Vautour retired from service on July 14, 1971, with a final salute of a formation of four Vautours flying in the skies over Israel.

On April 2, 1971, the first four A-4H Skyhawks arrived. The squadron operated the Skyhawks until the mid-1980's, when it was decided that 110 Squadron would become the second F-16 "Netz" Squadron in the Air Force.

Former Air Force commander Maj. Gen. *Dan Halutz* wrote in the squadron book: *"The heritage of the squadron is one of glory in the story of the Air Force. Training generations of pilots and technicians, making tens of thousands of sorties, including thousands of operational sorties in the Sinai "Kadesh" War, the Six Day War, the War of Attrition, the Yom Kippur War, the Peace of Galilee War and ongoing security operations between the wars."*

On September 5, 1980, the squadron was officially commissioned as the second F-16 "Netz" Squadron in the Israeli Air Force. On the day the squadron opened, it numbered six pilots: Lieutenant Colonel *Amir Nahumi*, commander of the squadron, major *Ehud Ben-Amitai* was first deputy commander, captain *Israel (Relik) Shapir* as second deputy commander, Captain *Shlomo Sas* acting as systems officer, Captain *Arnon Sharabi*, Captain *Dubi Ofer* and Major *Guy Shalev (Shelly)* as the squadron technical officer, with two F-16s (#100 and #111) on loan from 117 Squadron. At the request of the squadron commander, *Amir Nahumi*, the squadron's emblem was slightly changed from a pigeon carrying a bomb to a hawk carrying a missile.

January 20, 1981 - First squadron accident - Major *Ehud Ben Amitai* flying F-16A #222 collided during a training flight with an F-4E Kurnass with the same tail number. *Ehud Ben Amitai* was killed as well as captain *Danny Weiss*, the Phantom navigator in the back seat. The Phantom pilot managed to eject safely.

At the beginning of February 1981, the first conversion course in the squadron was started, in which nine pilots took part. The instructors were the first six pilots of the squadron. *Amir Nahumi* recalls: *"Simultaneously with the conversion course, we began training for long-range missions, in preparation for the attack on the nuclear reactor in Iraq. Already then, as a matter of fact some months earlier, Lieutenant Colonel Zeev Raz and I knew that we were going to face such an operation. The pilots themselves did not know what they were training for but that was all part of the plan."*

February 20, 1981 - *Udi Kramer* accounts: *"Back from a training flight with pilot Udi Peles as part of his conversion course onto the F-16, we landed long on a wet runway. Udi was unable to come to a full stop, but we were sure we would be able to stop without using the arresting cable. But when I took control, in the last few hundred meters the deceleration was not as expected, so we asked the tower to pop up the barricade. We made contact the barricade at a low speed, but not low enough to have been able to make it without it. Netz #017 entered the barricade and came to a halt. Damages were minor, not even the pitot tube was not broken".* An inspection of the plane revealed that there was a fault in the anti-skid system.

February 27, 1981 - The squadron goes on the first five minutes QRA (Quick Reaction Alert), which means it must be airborne within five minutes. On March 2, the squadron received its first mission, a 20,000-foot intercept patrol, but no contact was made. On April 23 two F-16s were launched for a patrol in the Tripoli area over Lebanon. On the way there they identified enemy aircraft, but no encounter ensued.

April 22, 1981 - World premiere attack for the F-16s. Objective: Concentrations of terrorists in the Sidon area. A formation of four F-16s with *Amir Nahumi* in the lead took off to attack the target. Control warned of MiGs in the vicinity, but no contact was made. The planes returned and the pilots reported good hits on the targets.

July 14, 1981 - The squadron writes history! Four F-16s took off that day on a patrol mission over Lebanon. Two F-16s patrolled the Tire area and the other two the Sidon area. *Amir Nahumi* (F-16A

F-16A #222 on its arrival day at the squadron on December 11, 1980.
Netz #222 collided during a training sortie with F-4E Kurnass #222 on January 20, 1981. ▶

▲ F-16B #017 accident on February 20, 1981. ▶

F-16A #219 - The world's first MiG Killer. ▶
(Photo: Moni Shafir)

37

#219) remembers: *"The Syrians tried to prevent our Skyhawks from attacking their designated ground targets. Shlomo Sas and I patrolled and waited for the Syrians, but they did not play ball. After more than an hour the fuel started to run out and I announced that I was going back to base. I turned towards home when on the way I heard over the radio that the MiGs were coming in for a fight. I turned back and started searching for them. I did not communicate via the radio, I just looked for them on the radar. It turned out that a pair of MiGs came towards us from the east, and Shlomo received instructions to intercept them. In the meantime, I locked on one of them, hoping they would turn south. One of them veered off to the north, while the other turned south as I had hoped for. I had a lock on him from 40 km distance. I dived and got him just where I wanted him to be. Everything was done at idle throttle. It was twilight and as soon as I got close to him, I suddenly saw that he had only one burner. I thought it maybe was our F-16, but I noticed Shlomo 6 km away from me. I launched a missile and the MiG exploded immediately. We started heading home when I announced on radio 'number one got a kill'. Suddenly everyone fell silent. That evening we threw a big party in the squadron. The date of the historic kill was very symbolic. Exactly 15 years to the day after a Mirage III downed a MiG for the first time".*

Shlomo Sas also tells: *"Between Squadrons 110 and 117, an unofficial but overt competition was in course over who would shoot down the first Syrian MiG. Spector, the Ramat-David base commander decided that since 110 Squadron was originally an attack squadron, while 117 Squadron was an intercept squadron, he would maintain that order ... Spector scheduled the 110 Squadron for the first attack of the "Netz" era and made sure to send the 117 Squadron on counter air patrols, as well as to place them on high alert, instead of the 110 Squadron. Thus, 117 Squadron would be scrambled first and kill as first.*

But the reality was different. True, the 117 Squadron shot down first…but their victims were helicopters (they do not count ...), The fact of the matter was that 110 Squadron was the first one to shoot down a MiG... This was on July 14, 1981, in a F-16A quartet formation with Nahumi leading, and Moshe Geffen, Shlomo Sas and Dubi Ofer as numbers 2, 3 and 4.

After a long patrol, the pair Nahumi/Geffen returned home due to lack of fuel, and Dubi and I continued to patrol.

Suddenly, the Northern control unit gave us directions to intercept two MiGs that flew west towards us, but before they reached us, they split up. One flew south towards Nahumi, and the other came towards us, a little to the south of us, completing a right turn from the west, north and then disengaging to the east. We engaged, me and Dubi, flying south, descending at high speed towards the MiG flying north.

I am locked, the lock is lost and then re-engages. The MiG is slightly to the right on the radar, west of me… Looking for a visual…

I planned to turn east over the left side, in order to close off the escape route for the MiG, when a call comes from Dubi, number 2, (who established eye contact with the MiG), asking me to turn right (because the MiG was on my right side). I turned right, over the long side, and because of the high speed, close to Mach 1, a large range opened up between me and the disengaging MiG to the east.

We started chasing fast at +Mach1, at low altitude. I am locked and waiting for the shoot cue in the HUD to flash - 'shoot'.

Waiting, waiting, waiting, but the Firing Cue did not come up. After a long chase, I noticed we were approaching the Syrian border, as well as an anti-aircraft gun firing at us. We got instructions from the controller to break contact.

I decided not to get the MiG and to break contact, turned my head, and cut off in a left turn, while pulling up the nose into a climb. I heard Nahumi shouting that he downed the MiG to the south ....

Returning home .... disappointed.

Landing at the squadron, friends from the squadron come and ask for my videotape. They run to the squadron HQ, leaving me with the mechanics to tell them what happened to us.

After giving them my account, I returned to the squadron offices and entered the briefing room. Everyone had their eyes turned down.

My video is being shown to me, with the pursuit of the MiG. And what do I see: when I pulled up the nose while breaking contact, with my head turned away, the firing cue begins to flash, the Firing Cue I had been waiting for to come up....

What to do, that is life. Later in the Lebanon war I got my kill…"

May 25, 1981 - As already mentioned, the tense atmosphere on the northern border continued and on May 25, 1982, during one of the scrambles, captain *Amos Mohar* (F-16A #240) shot down two MiG-21s.

Shlomo Sas, then the squadron's deputy commander, recalls: *"We were scrambled almost every day, but there were very few encounters with enemy aircraft. In between scrambles we continued with the training program. Up until the Lebanon war (Operation Peace for Galilee), everything was very tense. We sat for hours on standby alert, sometimes an hour and a half in the cockpit with the engine running, waiting to be scrambled".*

Mission orders continued to flow into the squadron operation room. On June 5, the day before the start of the Lebanon war, six formations of two F-16s took off to attack terrorist targets, followed by a formation of four, with later on a pair and another formation of four. Pilots waited in the pens or in the air for instructions.

June 6, 1982 Operation Peace for Galilee - On the opening day of the war, the squadron has 15 pilots

F-16A #240 - shows three Syrian kill marking on it.
Two MiG-21 downed by Amos Mohar on May 25,1982, and one Su-22 downed by Yehuda Bavli on June 11,1982. ▶

F-16A #242 - The Netz in which Dubi Ofer downed MiG-23 on June 8,1982. ▶

F-16A #250 - The Netz with which Avishai Kna'an downed MiG-23 on June 8, 1982. (Please note the fake Squadron emblem. Affixed to IAF planes when foreign visitors and photographers visited and the Air Force knew they might be published). ▶

F-16A #225 - With which Shlomo Sas downed MiG-23 on June 8,1982. Here on standby alert with its pilot in the cockpit with the engine running waiting for a scramble. ▶

39

and 19 armed and ready F-16s. During the five days of the operation, the squadron's F-16s took off on 283 operational sorties, most of them intercept missions. 19 enemy aircraft were shot down by 110 Squadron pilots.

Despite the amount of intercept patrols in the first two days the "kills" began to arrive only on the third day of combat.

Shlomo Sas accounts: *"During the Peace for Galilee War, on the night between June 6 and 7, the IDF planned to land forces at the Hawley River estuary. Prior to landing, "softening" of the landing area was necessary.*

110 Squadron was selected to carry out the mission. Accurate bombing was required, at night.

Prior to the arrival of the F-16 "Netz", the IAF did not

◄ *June 9, 1982 - Ofer Einav in F-16A #237 taxiing for take-off to the SAM missiles attack mission in which he shot down SAF MiG-21.*
(Photo: Lt. Col. Ofer Einav (res) collection).

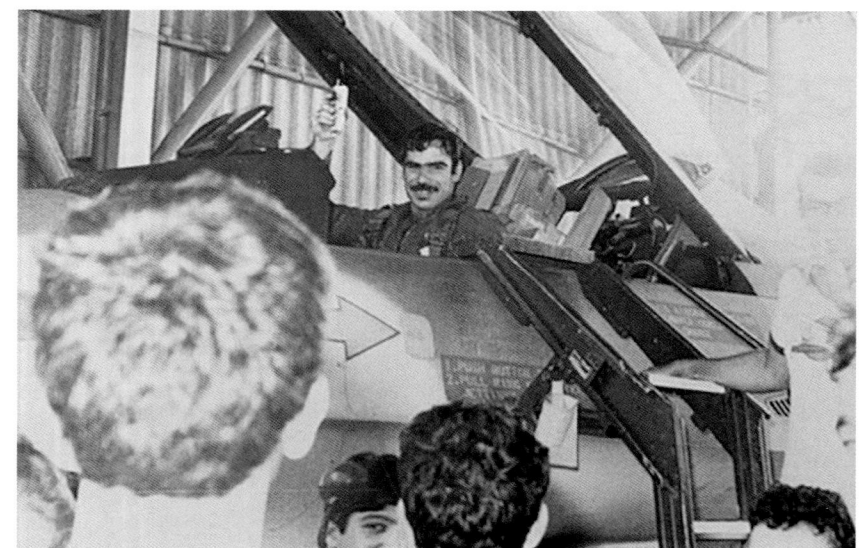

June 9, 1982 - Ofer Einav in F-16A #237 is received by the squadron ground crew after landing.
(Photo: Lt. Col. Ofer Einav (res) collection). ►

◄ *June 9, 1982 - Ofer Einav In the briefing room with the formation leader Relik Shafir.*
(Photo: Lt. Col. Ofer Einav (res) collection).

have an accurate bombing capability at night (there were no GPS bombs yet).

With the arrival of the "Netz" and its radar, a window of opportunity was opened for precision bombardment combined with a radar targeting. I was responsible for developing the doctrine of that warfare in collaboration with the special forces. And since I was the expert in the field, I was assigned to lead the operation.

I took off first with navigator (Moshe Barzilai) in an F-16B. The plan was to have the rest of the planes take off after I had bombed and made sure everything was going by the book.

In the field, despite the development of the warfare doctrine that lasted many months, things did not go smoothly ... The squad in the field erred in operating the equipment, and all attempts to rectify the situation were unsuccessful.

The fuel dwindled, I was getting close to bingo fuel. From the base they asked what is going on.... the navigator tells me: 'We are going back'.

I insisted. How could I go back? If I turned around, the other planes would not take off, and no softening would take place for the Israeli land assault...

I insisted.

Then just before I hit bingo fuel, I was able to find out what was wrong with the ground squad.

I instructed them on what to do, the equipment was straightened out, I bombed as planned! ... A sigh of relief, the stress had dissipated.

I sent all the planes into the air, they attacked.

The job was done, and the mission was a success".

June 8, 1982 - At 07:40 in the morning, a formation of four F-16s was scrambled led by squadron commander *Amir Nahumi* and with him *Dan Fredman*, *Yehuda Bavli* and *Avi Lavi*.

Upon their arrival at the scene, enemy MiGs were discovered on radar 26 km out. The formation performed the necessary maneuvers for the encounter, *Nahumi* identified one MiG in front of him and before he could lock on him, he discovered another MiG to his right. The MiGs turned north and the F-16s continued to chase them. During the chase, control reported two MiGs coming from the southeast at about 2 km distance. *Yehuda Bavli* and *Avi Lavi* immediately turned south to intercept them. *Nahumi* paused for a moment and when he realized that there was no chance to get a lock on the MiGs, he turned around. *Bavli* locked on a MiG that was coming across from the right and launched a missile. Everyone turned west without seeing the result. They returned to base without a kill, and as soon as they landed another formation was launched to intercept.

At 11:30 A.M., a formation led by *Shlomo Sas* with *Shmuel Gordon*, *Dubi Ofer* and *Avishai Kna'an* was scrambled. *Sas* and *Gordon* were vectored in by control for the intercept. They managed to lock on the MiGs, but they did not get permission to fire. On the other hand, *Ofer* (F-16A #242) and *Avishai* (F-16A #250) shot down a MiG-23 each.

Immediately afterwards, *Sas* and *Gordon* were directed to intercept the remaining MiGs. They spotted the MiGs and locked on to them. *Sas* thought that *Ofer* and *Avishai* were in front of him and ordered not to launch without visual identification. After a chase of about 25 seconds *Sas* (F-16A #225) launched a missile at the MiG-23 and it went up in flames immediately.

Shlomo Sas accounts: *"I was leading a four ship formation for patrol in the Galilee finger and over Lake Qarun, we patrolled in two pairs. Me as the leader, and Shmuel Gordon and Dubi Ofer and Avishai Kna'an the other pair. At one point we were directed to intercept MiGs flying south from Qarun Lake. I turned north with Gordon to engage, locked up, waiting for the other element (numbers 3 and 4), that was to the north of us, to pass us southbound, to get a "free field of fire". The bogey, a MiG-23, turned east in front of us, and disengaged in the direction of Syria.*

I went into a long chase after him (familiar...), trying to identify him before launching a missile. (Note ed.: a visual ID is always required before launching a missile). My F-16 is extremely fast, closes in on the MiG, and

F-16A #220 - With which Amir Nahumi downed a MiG-21 on June 9, 1982. ▶
(Photo: Sariel Stiller).

*almost approached gun range. ....
I do not have a visual ID yet. The system "shouts" at me - Shoot, Shoot...
At one point I see the MiG flying towards Syria, and realizing it is a Syrian MiG, I launched.
I pulled up the nose and climbed out, not looking at the flight of the missile (because you do not continue to fly straight for a long time, you can get "snatched").
I returned to base and performed a "kill buzz" around the control tower near the squadron.
A very slow and low " buzz ".
Everyone enjoyed it, consequently the base commander decided to ban "buzzes" after air victories for fear of safety issues ....
And that was the end of the count, a kill was achieved, this time the MiG did not get away".
There is another "story" to it:
"After landing, number 4 claimed to have downed the MiG. He also launched, without permission..., Apparently fed up with waiting. But luckily, he kept chasing the MiG, and in his video, you see my hit before his missile hits ....
On this mission, number 4 also shot down an airplane, launching without a go-ahead.
Dubi Ofer also shot down a MiG, but on the way home he had a power failure, the APU was activated, and he made an emergency landing (avoiding the "buzz")".*

June 9, 1982 - The fourth day of the Lebanon War. Seven victories over enemy aircraft in one day!
Ofer Einav remembers: *"At 15:28, four F-16s, a pair from our squadron and a pair from 117 Squadron, took off for a SAM missile battery attack mission (Operation Artzav 19 - Mole Cricket 19) in the Zavdani area in Syria.
No. 1 Relik Shapir - F-16 #223
No. 2 Ofer Einav - F-16 #237
No. 3 Rani Falk
No. 4 Sasha Levin
The targets were two SAM batteries, one pair of F-16s assigned to each. Ours was a SA-2 installation to take out, a little south from the SA-6 target of 117. On the way to the targets the skies were full of aircraft, participating in the ground battle in southern Lebanon and fending off Syrian intercepts.
The controller reported MiGs in the target area.
We did not meet them and attacked according to plan. At the end of the attack, the leading 117 pair announced eye contact with MiGs passing in front of them. Relik saw a MiG turning sharply in front of his nose in the direction of the other pair, he launched a missile that hit the target, the MiG went down in flames. I saw a second MiG passing almost head-on and starting a very sharp turn towards the other pair. I rolled in towards him, got on his tail and pickled off a missile. The MiG's tail separated and he spun wildly forward, wrapped in a gray cloud of fuel. Relik's MiG was slowly spinning, wingless and burning, the pilot ejected with a colorful parachute in the middle of the fur-ball with everybody watching. I passed close to him. At this point I disengaged to get back to my formation and was without eye contact, on my own. On the way west I spotted them far to the north and I continued westward alone. Heading out, I saw our other formations on their way to attack. An F-15 patrol turned in to me, taking me for a legitimate target coming Syria. Closing in, they identified me as friend and broke away, leaving me with my happy self. I spotted an F-16 formation above me on an intercept patrol, I pulled up and joined them west to the shoreline. From there I turned north and joined my formation on the way back home".*

A few minutes before *Reliks* formation landed, a formation led by *Amir Nahumi* (F-16A #220), with *Dan Fredman* in number 2, *Avi Lavi* (F-16A #255) as number 3 and *Roi Tamir* (F-16A #250) number 4, took off. After a while on patrol, control alerted for low-altitude MiG-21s heading southeast. *Nahumi* discovered a low-altitude MiG and locked on to it. *Avi Lavi* also locked on the same MiG, but as soon as he realized that *Nahumi* already got a lock on the MiG, he broke contact and turned into another MiG formation about five kilometers south. *Nahumi* continued to chase the MiG. It escaped, opening up a 6 km gap from the F-16 and began to turn north. The Israeli pilots increased speed and the distance narrowed. When *Nahumi* got within a range of 2.5 km, he locked on one of them and launched a missile. The MiG was hit and went up in flames immediately. He then turned in on another MiG, which was also chased by *Avi Lavi*, who asked *Nahumi* permission to take the shot. *Nahumi* confirmed and Lavi downed the MiG from

110 Squadron pilots at the end of the Lebanon War. Standing from right to left: Avi Lavi, Avishai Kna'an, Yehuda Bavli, Yoram Amitai, Dan Fredman, Moshe Gefen, Dadi Avner, Amos Mohar, Relik Shapir, Ofer Einav. Sitting from right to left: Gibori, Moti Rader, Roi Tamir, Gal Wissberg, Eyal Peked. Standing with his back to the photographer squadron commander Amir Nahumi. ▶
(Photo: Lt. Col. Ofer Einav (res) collection).

900 meters. *Nahumi* then ordered to break contact, get back into formation and check fuel levels. In that instant the pilots spotted MiGs firing missiles at *Dan Fredman*, who broke immediately and was unharmed. *Nahumi* and *Roi Tamir* approached the MiGs and *Tamir* locked on one of them, launched a missile that hit the MiG, which went down.

About an hour after the *Nahumi* formation had landed, another formation took off. The leader was *Relik Shafir* (F-16A #232), with him were *Avishai Kna'an* (F-16A #243), *Moshe Geffen* and *Moti Rader*.

A pair of low flying planes were discovered on radar. At some point the planes turned right. *Shafir* did not initially identify them as MiG's. However, at the last moment he did recognize them and launch a missile. *Avishai Kna'an*, who was already locked on the second MiG, immediately took the shot, and downed him.

A happy day in the squadron. Seven MiG 21s in one day. It was the best day of air battle the squadron has known to this day.

June 10, 1982 - On this day, the squadrons fighters constantly set out on patrol missions. One of the elements that took off was *Amir Nahumi* (F-16A #237) and *Shmuel Gordon*. Their mission was to patrol the Mount Hermon area. Once airborne, they were directed by control to patrol the Qaraoun Lake area. *Nahumi* spotted enemy aircraft over the southern part of the lake and as he approached, he identified a pair of MiG-23's. A long chase ensued, as the MiG's increased speed and ran away. The lead MiG lost control, entered the wadi, and crashed. *Nahumi* locked on the second MiG, launched a missile, and downed it.

A few minutes later while in the air, a message was received about more MiG's. *Nahumi* again locked from a range of 1,500 meters and intended to launch a missile. *Gordon* asked *Nahumi* to grant him the kill. *Nahumi* acknowledged, but because of sharp evasive maneuvers by the MiG-21 *Gordon* was unable to down him. *Nahumi* locked the MiG again, launched a missile, job done.

June 11, 1982 - In the morning, a formation led by *Relik Shafir* (F-16A #225), with as number 2 *Yehuda Bavli* (F-16A #240), number 3 *Dubi Ofer* and *Avishai Kna'an* in number 4, took off for a mission of protecting the area over Lake Qaraoun. Control gave the alert for enemy aircraft. *Relik* locked on two Su-22s west of the lake. He launched a missile at one of them and ordered *Bavli* to launch at the other. Both Su-22s were shot down. Then a message was received about more MiGs. *Shafir* blocked the escape route for the enemy planes and then identified another pair of MiGs. He managed to shoot down one of them, when control ordered them to break contact and return because the area was protected by SAM missiles.

At about 09:16, another formation took off led by *Amir Nahumi* (F-16A #237), with *Roi Tamir* (F-16A #246) as nr. 2, *Moti Rader* in nr. 3 and *Avi Lavi* in nr. 4, heading towards the same patrol area. Upon arrival, they were alerted on MiGs coming in from the east. The pilots got in formation to protect the area while *Nahumi* searched for the MiGs. He discovered four of them and fired a missile at the rear MiG that was immediately hit and crashed. He instantly continued in pursuit of the nr. 3 in the MiG formation and knocked him down as well. Before getting on the tail of the third MiG, *Nahumi* checked with *Tamir* about his location and noted that the latter was also locked on the same MiG. *Tamir* launched and got the kill. The fourth MiG ran off alone, entered the wadi (dry riverbed), and disappeared behind a ridge.

During the operation, the squadron made dozens of operational sorties within five busy days. Despite the tense atmosphere and pressure, no aircraft were hit or damaged, and not one single mission was canceled. The squadron's ground crew worked around the clock and there were no technical glitches.

The "Netz" had proven to be an excellent fighter with outstanding weapon systems.

March, 1983 - *Lt. Col. Amir Nahumi*, handed over command of the squadron to *Lt. Col. Naftali Maymon*.

August 28, 1983 - As part of a battle practice day at the Ramat David AB, a formation of four F-16s took off for a training mission. The target was around Ramon AB. Squadron Commander *Naftali Maimon* led the formation. The weather was great. Number four in the formation was Captain *Dan Fredman* in Netz #227. Getting out from a diving attack, the jet hit the ground and crashed.

June, 1985 - *Lt.Col. Naftali Maymon*, handed over command of the squadron to *Lt. Col. Ofer Lapidot*.

April 10, 1986 - Netz #240 took off for an engine test flight. At an altitude of 4,500 feet and a speed of 170 knots, its engine shut down due to failure of a fuel regulator. The aircraft was at that time around the Elyakim Junction. The pilot *Eyal Peled* tried to restart the engine but without success. *Eyal* slid in the direction of Yokneam. He realized that he would not be able to make it back to base for a landing, directed the plane to an uninhabited area and ejected at an altitude of 250 feet.

May 7, 1987 - *Lt. Col. Ofer Lapidot*, handed over command of the Squadron to *Lt. Col. Shlomo Sas*.

July, 1987 - *Lt. Col. Shlomo Sas*, in fact, took command of the squadron in the final stages as a "Netz" squadron and prepared it for the more advanced F-16C jets. The squadron's F-16As and Bs were transferred in mid-July to the squadrons at Ramon AB.

PHOTO GALLERY

◀ F-16A #261 at Fort Worth's General Dynamics plant checked over before shipment to Israel - June 10, 1981.
(Photo: Major E.G. collection).

◀ U.S. Air Force pilot after a long ferry flight to Israel. Received with a bouquet of flowers and a chilled bottle of beer by the squadron crew (From left to right: adjutant officer Orly Tzur, maintenance officer Zvi Agassi and technician Yair Tal).
(Photo: Major E.G. collection).

◀ F-16A #110 In honor of one of the squadron's anniversary, the number 110 was painted on one of the squadrons' F-16As. (Photo: Lt. Col. Ofer Einav (res) collection).

F-16A #227 which arrived at the squadron on January 21, 1981, is seen here returning from a training sortie, with its original factory tail number. ▶

F-16A #228 which arrived at the squadron on January 21, 1981 taxiing to the take-off point. ▶

110 Squadron deployed to Hatzor AB (1981) for a series of air-to-air combat training.
(Photo: Lt. Col. Ofer Einav (res) collection). ▶

▲ *F-16A #237 carrying its three kill marks taxiing for take-off.*

▲ *F-16A #248 awaiting its pilot, ready for the next training sortie.*

◀ *F-16A #248 with its factory applied tail number during landing at Ramat-David AB.*

▲ *F-16B #015 was one of the two-seater aircraft used to convert and train new F-16 pilots.*
(Photo: Lt. Col. Ofer Einav (res) collection).

▼ **110 Squadron kill Board.**

OPERATION OPERA
Strike on the nuclear reactor in Iraq
by Tsahi Ben-Ami

"In the early evening of Sunday, June 7, 1981, the quiet peace that prevailed in the historic city of Baghdad on the shores of the river Tigris was suddenly and brutally disturbed by the sounds of intensive bombardments. The messengers of death and destruction had struck once again. The Israeli Air Force was engaged in an armed attack deep into the territory of Iraq. Their target this time was Iraq's nuclear research station at Tuwaitha just outside the city of Baghdad. The attack resulted in the loss of many human lives and extensive material damage.
...This incident has given rise to the present complaint by Iraq against Israel. As it happens, the facts of the case are not in dispute. What is in dispute is Israel's claim that its raid on Osirak falls within the purview of the doctrine of self-defense... according to the Israeli mythology of self-defense-it can only lead, as once before in history, to national suicide, except that this time it may be so apocalyptic that the fate of all of humanity will be equally imperiled."

(2282nd MEETING of the UN Security Council, NY, Monday, 15 June 1981)

Twenty years later, at a reception in Washington, US president *George W. Bush* presented *David Ivri* - Israel's ambassador to the United States and previously the IDF/AF Commander, as the one responsible for the raid on the reactor in Iraq. In perspective of twenty years, and ten years after "Desert Storm" – President Bush admitted: *"If you would not have attacked the reactor at that time – we would have been in a completely different situation..."*

The following lines summarize the events, decisions and cockpit impressions from the Israeli raid on the reactor in Iraq – "Operation Opera".

As surprising as it sounds - The Iraqi Atomic Energy Commission was established with U.S. help and encouragement in 1956, as the United States donated most of the unclassified reports of the U.S. Atomic Energy Commission (AEC) and provided training for the first generation of Iraqi nuclear scientists.
Following the military coup of 1958, the Soviet Union strengthened its relations with Iraq, and in 1960 provided with Iraq a small research reactor and other support facilities – all located at the Al-Tuwaitha site, about 30 kilometers south of Baghdad – turning the site into Iraq's Nuclear Research Center.
Between 1972 and 1976, Iraq's nuclear weapons program was started. Its initial goal was to acquire a complete, safeguarded fuel cycle able to produce separated plutonium. For this purpose Iraq secretly signed an agreement with the French government to supply a 40 MW MTR reactor called Tammuz-1 or Osirak, a zero-power reactor called Tammuz-2, a material testing hot laboratory (named LAMA), workshops, and a radioactive waste treatment station (RWTS). In addition, Italy supplied a research-scale radiochemistry laboratory capable of handling plutonium and an experimental fuel fabrication laboratory. The two French supplied reactors were housed in building 24.

The other major building erected by the French was the LAMA facility in building 22. It housed facilities for fuel testing and preprocessing of spent fuel before it would be sent to the reprocessing facility or storage.

During the 1970s and 1980s, the Tuwaitha Nuclear Research Commander Center underwent a dramatic growth, much of which involved new buildings and facilities that housed clandestine nuclear weapons-related activities in violation of Iraq's commitments under the Non-Proliferation Treaty (NPT).
The International Atomic Energy Agency (IAEA), responsible for verifying that Iraq complied with the NPT, was ignorant of the building activity at Al-Tuwaitha, although this large-scale construction activity in areas well removed from the known civil facilities was visible to U.S. and Russian intelligence satellites.

The Israeli intelligence became interested in the Iraqi nuclear program in March 1975. Information was 'flowing in,' said former head of Ama'n (Agaf Modi'in – IDF Intelligence Branch), General *Yehoshu'a Sagi*, as he described the Intelligence gathering procedure concerning the 'Osiraq' project. (Osiraq – combining the names Osiris and Iraq – was the name of the reactor Iraq bought from France. This project was also known as 'Tamuz 17' - "Tamuz" was the name of the Babylonian god of hell, as well as the Babylonian name for the month of July. Tamuz 17 (July 17th) was the date in which Iraq's "Ba'ath" party took over rule in Iraq.) Three years later, in August 1978, the subject was brought before the Israeli government.

General *Aharon Yariv*, former head of Ama'n, was appointed by Prime Minister *Menahem Begin* to assess the Iraqi nuclear program. *Yariv's* report confirmed Israel's worst nightmare - Iraq was looking for strategic capabilities! According to *Yariv's* report and his experts, the French reactor and Italian labs would provide Iraq with the capability of producing as many as four nuclear devices every year. *Yariv* also warned against attacking a "hot" reactor (a reactor loaded with nuclear fuel – such an attack, depending on wind direction, could cause more than 150,000 casualties in Iraq.

Israel under Foreign Minister *Moshe Dayan* launched a diplomatic assault against France and Italy in an attempt to prevent them from cooperating with Iraq. France, under Iraqi pressure, rejected Israel's suggestion to provide Iraq with "Caramel" – a degraded nuclear fuel that would still enable Iraq to conduct nuclear research, but deny it from generating fissile material used for nuclear devices. As diplomacy failed, *PM Begin* appointed *Nahum Admoni*, the Mossad's second in command, as head of the "Idan Hadash" (=New Era) group "Idan Hadash' was a combined team of Mossad, Ama'n and other professional bodies and was assigned to come up with ways to confront the Iraqi threat.

Saddam Hussein held a speech on October 29th, 1979, in which "The struggle with Israel will be long and hard, and Israel might even use atom bombs against the Arabs. That's why the Arabs should prepare the required means for victory" and further, his interpretation of Jihad as: "The law of Muhammad lies in the sword, while the law of Saddam lies in the atom." – Made it clear that Iraq's nuclear program posed a threat to Israel's existence.

In 1978, Israeli Defense Minister *Ezer Weizman* challenged IDF/AF Commander Major General *David Ivri*, by asking him to prepare a plan for striking the reactor in Iraq. It was a challenge indeed; *David Ivri* and the Head of AF operations, Brigadier General *Aviem Sela*, had to come up with both technical and operational solutions. Al-Tuwaitha was beyond the operational range of Israeli fighters; flight would be mostly over hostile territory with pilots having to avoid detection, and the worst scenario – if shot down - pilots need to be recovered. Another issue of concern was the inability to predict the weather along the route and at the target area.

Feasibility studies were initiated to test the ability of carrying out the mission using the F-4E "Kurnass" (Phantom). Since the reactor was out of the Phantom's range, the fighter had to be refueled by special tankers equipped with a boom system. These tankers, however, were denied to Israel as they were considered a strategic system. Since Israel was not refused the 'probe and drogue' refueling technology, the IDF/AF began installing in-flight refueling probes on its Phantoms, enabling them to be refueled by either KC-130s or A-4Ns "Ayit" (Skyhawk) equipped with the D-704 "Buddy refueling" pods.

As soon as the fighters were modified, feasibility studies continued. Under direct orders by *Sela*, pilots were sent out on bizarre training sorties in which they were asked to rendezvous with A-4 "tankers" while maintaining complete radio silence and using nothing but passive navigation techniques, without being assisted by regional ground control units. Later, pilots were asked to do the same in various weapon configurations, ending the flights within bombing ranges. Pilots were constantly asked to report fuel consumption, but actually none was told what they were training for. In February 1980, *David Ivri* told *Weizman*: "it can be done".

The idea of *Ivri* and *Sela* was to use a combined force of Phantoms refueled by Skyhawks. "It was a complicated operation in which only four fighters, out of a force of 18 or 24 aircraft, would actually reach the target with enough fuel to carry out the strike mission. One can imagine the risk involved in replacing the large strike force with slow KC-130s flying over hostile airspace".

During that time, in 1979 a mysterious explosion (alleged by foreign sources to be the work of the Israeli Mossad) destroyed the reactor's core while awaiting shipment in France.

A significant development was about to change the plans. *Ivri* explains: *"In 1979, US Defense Minister, Harold Brown, came on a visit to the Middle East. While meeting Ezer Weizman, Brown offered him 75 F-16 fighters originally ordered by Iran (The deal was canceled due to the Islamic revolution and U.S. hostage crisis.) Ezer called me to his office and asked me if I would accept the Iranian fighters. I said yes. With all the difficulties involved – if somebody gives – I'll take!"*
Brown's offer enabled a significant reduction of the pilots at risk – *"That's why I jumped at the opportunity of getting those F-16s. Then,"* continues *Ivri, "as I sent the pilots for the F-16 conversion in the US, I told them to check how far they can get on strike missions. The pilots were shocked! Their vision of the F-16 was of an agile, potent interceptor and the first to engage in air combat – they did not think of it as a bomber. Pilots tend to favor air-combat over long range air-to-ground raids!"*
"I told them…" continues *Ivri, "you'll have plenty of dogfights, but I need this 'thing' too, and then I made sure that we'll get plenty of drop tanks for the new fighters."*

At Ramat David AFB, July the 2nd was not an ordinary day. People everywhere were just standing, staring at the sky, following the trail of smoke of a Phantom closing-in while escorting four smaller dots - F-16s!

Tension along the northern border was high. Terrorists were attacking Israel, and Syria deployed SAM batteries in Lebanon – sheltering the terrorist bases.

Once more *Sela* initiated his feasibility studies, sending pilots on missions asking them to report fuel consumption in various weapon and flight configurations. *"I thought we were training for yet another futuristic war scenario. Tension has built up in the north and it seemed a reasonable assumption,"* later explained Lieutenant Colonel *Ze'ev Raz*, squadron leader of the 'First Jet Squadron'. Then, one day, *Sela* summoned *Raz* and the squadron's young navigation officer, *Ilan Ramon*, for a meeting with General *David Ivri*. Ivri pointed his finger at the map showing Baghdad and said - *"You will strike here…"*

* * *

"A big clock stands over us, and it is ticking away. The fact, that Iraq is developing nuclear weapons, means danger for every man and woman in the state of Israel. Saddam Hussein will not hesitate to use such a mass destruction weapon against us…"(Prime Minister *Menahem Begin*).
It was at that meeting, on October 29th, 1980, and after more then two years of debates, that the Israeli government decided to attack the reactor in Iraq.

* * *

The situation in the region was not a contributing factor. Iran attacked the reactor twice without causing any significant damage, but caused Iraq to tighten the reactor's air defense, which now included balloons against low-flying aircraft, a high and wide dirt wall and an array of SA-2, SA-3, SA-4 and SA-7 "Strela" missiles. Tension on the Israeli northern border reached another climax when Israeli F-16s shot down two Syrian Mi-8 helicopters on April 28th, 1981. Iraq and Syria were both on high alert, and the Israeli AF was busy fulfilling the Israeli-Egyptian peace agreement while withdrawing from Sinai.

Pilots were assigned for the mission based upon their skills and combat experience. In an interview with the IDF/AF magazine Brigadier General *Amos Yadlin*, now Chief of IDF/AF Staff Directorate said: *"It was clear to all of us that only a very special operation requires that extraordinary group of pilots. It is not common that a squadron's deputy commander gets assigned as a formation's 'number two'".*
"There was quite some irony in the fact that pilots were training for a mission without knowing what the mission was; we were just told that we had to reach full operational capability by October".
"When I was asked to prepare a flight plan that would enable the F-16 to reach an operational radius of 600 miles, I took a map and a ruler and drew a circle with Ramat-David as a center. That was the first time I understood the idea of the planners. The only worthy "Arab" target at that range was Baghdad".

Then the squadrons began training for the mission, concentrating on attacking infrastructure targets and testing the capability of the F-16 to reach the planned range. At that point, another technical issue popped up: The F-16s manual forbids dropping external fuel tanks while the aircraft carries its bomb load. Depending on their manufacturer, on release the tanks may hit the bombs. A test at Manat – (Merkaz Nisui Tissa - the IDF/AF Flight Test Center) actually proved the opposite – *"with the bombs mounted on the wings – the tanks fall off in a very smooth fashion".*

Training flights continued. *"We spent quite a considerable time in training all sorts of evasive maneuvers that would help evade interception. We also trained in engaging air-combat without lighting the afterburner – after all we were very short on fuel."* According to the flight plan, ETA was set for the late afternoon, providing enough light for the pilots to aim at the target, but also plenty of dark hours in case a recovery operation would be required. *"We all knew everything would be done to rescue us, however it was clear to us that a chance for safe recovery from within the 'last 100 mile zone' is extremely low,"* said *Yadlin*. Former IDF/AF commander, *Avihu Bin-Nun* revealed recently that rescue helicopters were deployed in Saudi-Arabia for that purpose, along the Iraqi border.

"We had to set D-day for mid-1981, avoiding Sadat's visit to Jerusalem (June 4th) and Begin's visit to Egypt (August 1981). We decided the attack would take place on a Sunday, assuming the French consultants would not be working" – explains *Ivri*.

D-day was set for Sunday, May 10th, 1981. Pilots from the 'First Jet Squadron', 'Knights of the North' and 'Knights of the Double Tail' squadrons were ordered to remain at their squadrons.

Captain *R.*, an F-16 pilot, recorded the preparations for the strike in a diary: *"Sunday, May 8th, 1981. We arrived at the squadron quarters. On "final" we see our "twin tailed" friends – some were already taxiing down the runway – joining us for the joint briefing. The briefing room was crowded with high ranking officers – Major General Refa'el Eitan – the Chief of Staff, IDF/AF Commander – David Ivri, and many more officers and officials.
Ze'ev Raz began the briefing – take-off, navigation, threats, the rivers and the reactor's wall on the*

horizon... "The future of our state depends on you," concludes Captain *R*.:
"*I return to the squadron, putting on my g-suit, verify that my pistol is loaded when suddenly the speaker screams 'everybody back to the briefing room'... the mission had been aborted...*"

Some of the pilots had already climbed into the cockpits, when a phone call came from the Prime Minister. It turned out that the head of the opposition party, *Shimon Peres*, called *Begin*, warning him of the possible political repercussions for the attack. *Begin* was worried about the fact that *Peres* knew about the raid. If *Peres* knew – so could the enemy.

"*Personally, I thought the government wouldn't dare carry out such an operation,*" says *Yadlin*. "*As someone who participated in the planning, training and preparations of 'exotic' operations, I lived to see many of them end up in a drawer – I was sure this operation was about to join that long list.*"

On Thursday, June 4th, in his office, Colonel *Yiftach Spector*, air wing commander of the Ramat David AB, *Ze'ev Raz* commander of the 'First Jet Squadron', and *Amir Nahumi*, commander of the newly formed 'Knights of the North' squadron, got the message - "Operation Opera" was just about to take place.

Ivri describes the fighter's configuration: "*We tried (during the feasibility studies) to launch aircraft with 12 bombs mounted on TERs (Triple Ejector Rack), but this configuration increased drags and fuel consumption. The most drag efficient configuration included 3 drop tanks (one on the centerline and two under the wings), two AIM-9 missiles mounted on the wingtips and two Mk-84 dumb 'iron' bombs filled with Tritonal explosive material*". As for the weapon selection: "*We used simple GP 'iron' bombs – the simplest ordnance to minimize TCT and resulting exposure to the SAM batteries. We had no time to guide the bombs, and this concept proved itself. Not always the most sophisticated solution is the best solution. In this case we made a simple but perfect match. Using the delayed fuses enabled us to be very accurate.*"

The strike force included 8 F-16s and 6 F-15s – equipped with CFTs – providing air cover, ECM cover,

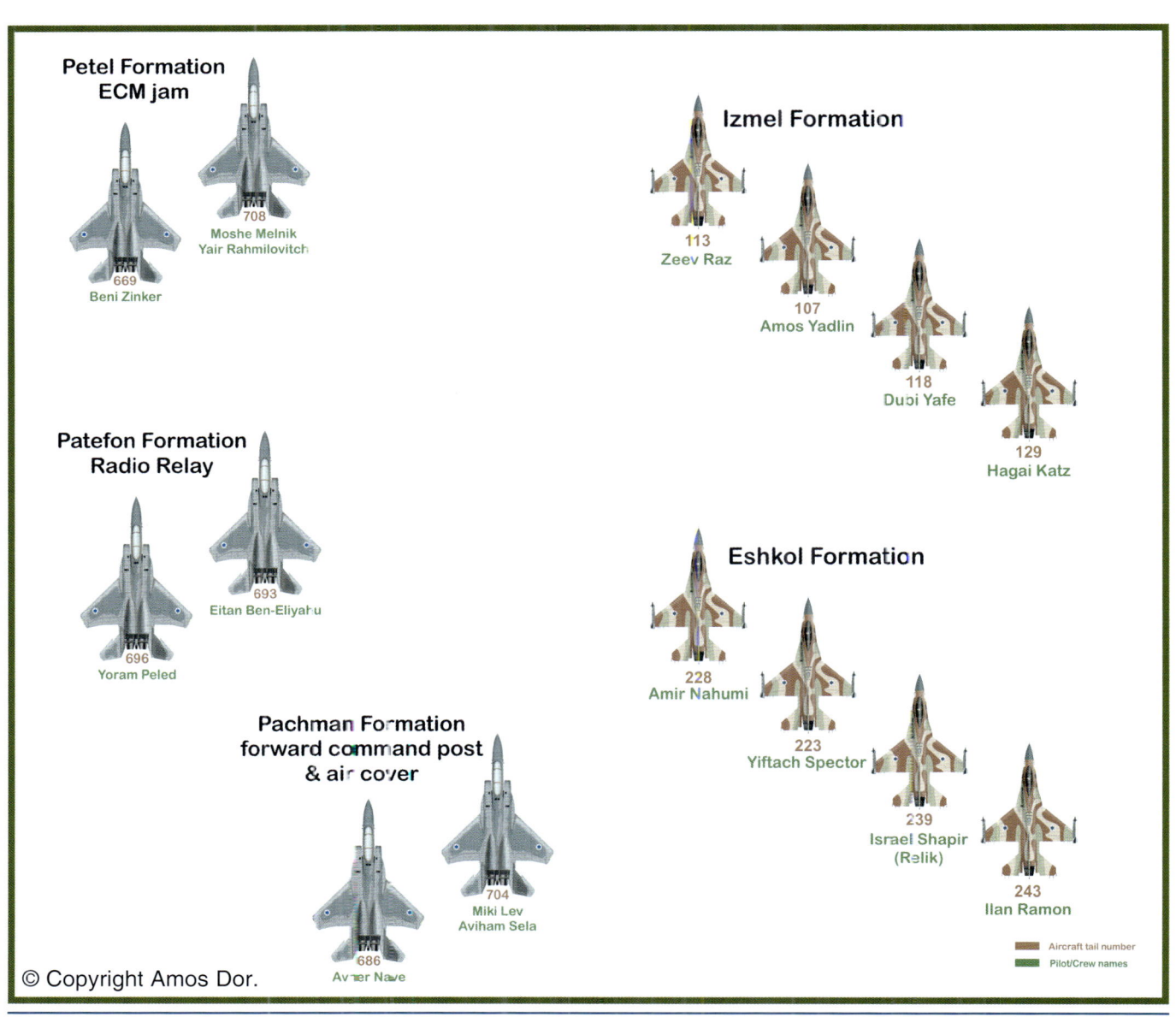

▲ *"Izmel" formation before takeoff.*

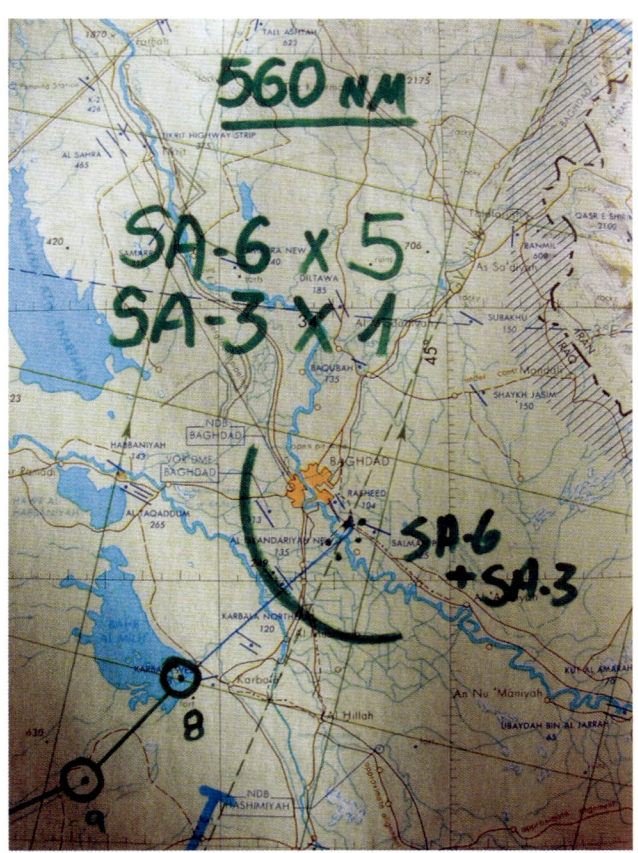

Target map.

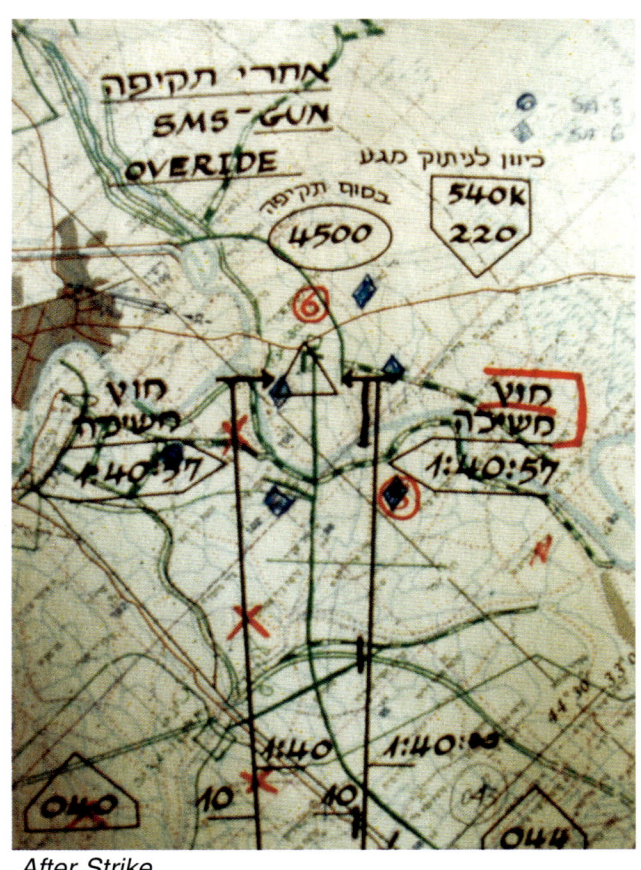

After Strike.

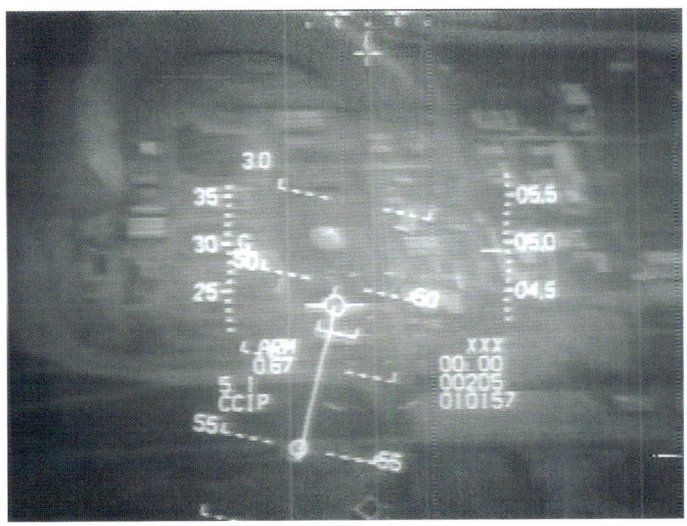
Target HUD view.

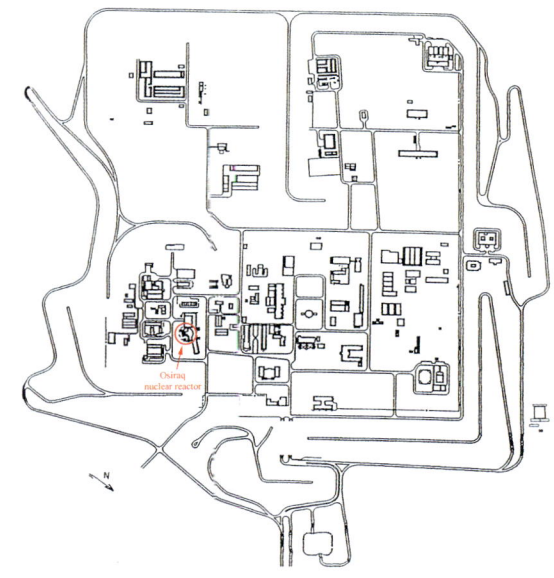
Map of the reactor site.

The target before ▲ and after ▼ the strike.

and radio relay.

R. describes: "The briefing came to an end, and one of the pilots began handing out dates to his fellow pilots. With a smile on his face he said: "You should get used to Iraqi food…"
I go to the restroom, but don't drink any water so as not to burden on my bladder on such a long flight. I grab my helmet, G-suit and VTR cassette and go to my F-16. I take a look at that great looking fighter, and think to myself – this is our fourth flight together – don't betray me this time" and pet the aircraft by sliding my hand over the AOA sensor, I pound on the bomb and climb up the ladder to the cockpit.
I start the jet, and move the throttle to idle. The jet catches up and temperature climbs to 500 degrees – "we have several hours to spend together, metal and flesh" – I look around and see one of the pilots running around worried. His F-16 developed a problem and he had to replace the fighter. Thirty minutes for calibration. I roll outside a bit, the aircraft is heavy and the sun is burning hot. I continue rolling – taxiing to the refueling position, open the canopy and wait for the fuel indicator to rise. Everybody around gets loaded with fuel, but my F-16 does not seem to care. I ask the mechanic to replace the bowser, but there seems to be no change. OK. I take the call and decide to go on – I have enough fuel. 15:55 - we line up for take-off, maintaining a safety gap of 300 meters between each other. 16:01 – Time for take-off - I began acceleration, watching digits change on the HUD; full dry power – taking-off like a Fouga Magister. 80 knots; afterburner on; my F-16 is heavy and it is fast; 190 knots – afterburner off – back to full dry power; I retract the gear, gain altitude and take my position in the defensive formation. Initially, at 500ft., The view is breathtaking but then it all changes into dunes and empty desert scenes. I keep my place in the formation. We go down to 150 ft. Time seems to stand still, and I seem to use more fuel than planned. I pump some fuel into the rear tanks to reduce the drag, it might help…"

Yadlin: "We flew over the south of Jordan and headed into the Arabian Desert. It's a wide, empty and very desolated area that posed a serious navigational challenge. Then we entered Saudi Arabia. I see some military bases, paved roads – many of which were not indicated, on my map. As we penetrated Iraq, a medieval fortress caught my eye. It was very large and in ruins. If we knew about it – we would have used it as a way-point."

The strike force reached its fourth way-point Bahr Al-Mil. Yadlin continues his description: "Bahr Al-Mil is a very large lake, located about 50 km west of Baghdad. The map clearly shows an island - a long strip of land – in the lake, but instead we saw four little islands. We were right on spot, but things were not going as expected… Raz took a courageous decision and continued as planned. Later it turned out that we had arrived after a very rainy season which had covered parts of the island".

"We flew over a green, inhabited area. Raz broke the radio silence warning us of rising power lines. We were forced to climb, exposing ourselves for the first time to the Iraqi radar. We accelerated to 540 knots, flying within the SAM kill zone. Then, on the horizon I saw the high dirt wall surrounding the reactor and the balloon defense system we were warned about".

"I do another check-up, I switch my hands and hit the switch – tack… both drop tanks fall off the wings – what a relief… Another way-point, and the view changes – 'Iraq-land'! As soon as I dropped the drag generating fuel tanks - the aircraft became much more fuel-efficient. Actually it seemed as if the fighter began generating fuel – I had much more fuel then calculated, my fuel reserves grew as we closed in on the target. An occasional truck crosses the desert, and I can see the lake with four islands. The way-point designation changes and we accelerate. Another longitude is crossed. We break the formation and line up.
We cross the shore, on the right – a bus full of tourists; colorful taxis stand at the beach – people having fun. Four miles before pull-up. I turn on the radar, and see no MiGs – I pray that the scope will remain empty, and then from left to right I see this long blue river – the Euphrates! I think to myself – we've been here before – 2,500 years ago…" R. continues: "Buildings pop up. An old airport, power lines – everything passes by at a vast speed. Ten miles to go – I feel my bladder. A flash in the air and Raz breaks the radio silence: "pay attention to ground fire" – chaos begins.
Flashes fill the sky, followed by little puffs of smoke. Another second and I see the Tigris – I push the throttle forwards - full afterburner. I pull up and see the wall and then bright in the dusk light I see that reactor's silver dome. It is 17:35, 6,500ft. and 7g – flying upside down - gaining speed; I see some dust rising, but no explosions. It's just me and the pipper. AAA fire bursts out from the compound, while the pipper slowly follows the trajectory line. Make no mistake. Another small correction and that's it! 3,500 ft. One pickle and the bombs fall off. I break hard to the left and look for missiles in the air. One missile zooms below me and gets stuck in the ground. 600 knots, full dry power – I climb 40,000ft. and join the formation."

"I felt that thud on the airframe as the bombs were released" describes Yadlin: "and got out of there as fast as I could. That's the great advantage of 'iron' bombs – you don't have to guide them to the target - all you need to do is drop the bombs, find your formation, protect your wingman, protect the formation and verify that nobody took the opportunity to get on your tail". The fact that the formation flew back into the sunset was another contributing factor to the strike force's safety. Any intercepting Iraqi pilot, chasing the Israeli strike force, would fly with the sun in his eyes – a fact that will make identification difficult for both pilot and heat seeking missiles.

It was time for count down. Raz asked the pilots to call in, but there was no reply from Eshkol 4. Eshkol 4 flown by Ilan Ramon was the last pilot to attack the reactor – putting him at the greatest risk. Ramon was also the most inexperienced pilot in the formation – flying his first combat mission on an F-16… Raz made another attempt: "Eshkol 4 – are you OK? Ramon came in on the radio: "OK, OK I pulled out of it OK!" - Ramon was tied up evading the ground fire – too busy to respond…
Relieved - Raz called in on Sela:
– "Pachman from Izmel 1. Izmel and Eshkol are all Charlie".
Charlie was the code name for full success.
Sela still refused to believe:
– "Are all Charlie?"
– "Affirmative all Charlie! Affirmative! Target destroyed, two bombs overshot."
– "Any damage?"
– "Maybe some scratches from AAA – nothing serious".

"On the way back the pilots flew as high as possible as this was the most fuel efficient option." Explains Ivri – "It is obvious that the fighters were exposed to enemy radar – so we kept flying out of the intercept range of enemy air bases, using the F-15 escort as a shield. There was just one threat – a single Iraqi MiG that closed in, but did not engage. We were worried about the Jordanian AF - The late King Hussein spotted the Israeli fighters while cruising on his yacht in the Bay of Aqaba. Hussein made some calls to Saudi Arabia, but the Saudis saw nothing. It was true. The Israeli fighters flew as one in complete radio and electronic silence at a very low altitude."

"The first time I spoke to the pilots, was just minutes before the landing" said Ivri – " At the end of a major mission - tension fades away, and the pilots get joyful about their success – that's when they get careless. I took the microphone and called Raz: "Parpar to Izmel – remember! Missions end only in the hangar!" – I wanted to make clear that the mission was not over yet."

A year later, as the participants of the raid gathered, commemorating the raid on Iraq, Yiftach Spector, who overshot the target, brought a champaign bottle. On the label he wrote: "This raid proves how seven eighths make 100%…"
Information gathered after the Gulf War made the world realize the devastating capability of Iraq. Several years ago, with this new intelligence - Ya'akov Amidror – an Ama'n officer, initiated a retrospective research to see if the new information still justified the Israeli raid. The new documents proved the necessity of the strike on the reactor.

Participants in Operation Opera in the 140 Squadron building in Etzion AB after returning from Iraq.
Standing from right to left: Israeli Shapir (Relik), Shlomo Sas and Rani Falk (The alternate pilots), Amos Yadlin, Dubi Yafe, Amos Mohar (110 Squadron Operations officer), Iftach Spector.
Sitting from right to left: 140 Squadron Commander Naftali Maymon, Zeev Raz, Ilan Ramon, Amir Nahumi, Hagai Katz and Dubi Ofer.

 # The Negev Squadron

The 253 Negev Squadron was established on August 1, 1976 at Hatzor AB in order to take over the IAI Nesher fighter jets from 113 Squadron that started to operate the IAI Kfir fighter jets. On December 1st that same year they moved to the newly constructed Eitam AB. At one point the IAI Neshers were sold to Argentina and the squadron was converted to operate the Mirage III - "Shahak". With the signing of the purchase contract for the F-16 fighter jets, the Air Force decided that the Negev Squadron would become the third "Netz" squadron, and that her new home would be Ramon AB, at that time still under construction in the Negev. *Lt. Col. Gideon Livni* was appointed squadron commander, *major Amos Yadlin* and *captain Ilan Ramon* were designated as his deputies, and *major Dov Baror* took the job of technical officer.

In September 1981, the squadron opened as an independent unit within 110 Squadron at Ramat David AB.

The pilots flew and trained on the 110 Squadron planes. On February 25, 1982, the squadron left Ramat David AB and in a big ceremony moved to Ramon AB, flying its 16 F-16s with the IAF Commander *David Ivri* in the back seat of one of the F-16B aircraft.

June 6-11, 1982 Operation Peace for Galilee
During the First Lebanon War, the squadron conducted 192 intercept and escort patrol missions. Its pilots shot down five enemy aircraft within the same day.

June 9, 1982 - Two formations took off at almost the same time, the first led by *Gideon Livni* with *Moshe Rosenfeld* (F-16A #272), *Amnon Sharabi* (F-16A #287) and *Ofer Safra*. *Gideon Livni* experienced a main landing gear problem, had to turn back to base and *Ofer Safra* took the lead. The second formation was led by *Ilan Ramon* with *Uri Gil* (F-16A #290), *Amir Ivtzen* and *Nimrod Gil* (F-16A #267). The first formation encountered MiGs and engaged them. *Safra* launched a missile on the MiG-23 but it did not hit. *Sharabi* and *Rosenfeld* launched on the same MiG together and it was shot down. Each of them was credited with a half kill. *Rosenfeld* then locked on a MiG-21, launched and downed him.

The second formation also encountered MiGs. *Nimrod Gur* (F-16A #267) downed one MiG-23 and *Uri Gil* (F-16A #290) downed one MiG-21 and one MiG-23.

May 1983 - *Lt. Col. Gideon Livni* handed over command of the Squadron to *Lt. Col. Avi Barber*.

May 1985 - *Lt. Col. Avi Barber* passed command of the Squadron to *Lt. Col. Yossi Eliel*.

July 3, 1986 - *Lt. Col. Yossi Eliel*, commander of the Negev Squadron, was training aerobatics at low altitude flying in F-16A #277. He practiced for the Air Force Day air show later that month. While performing a low-altitude loop above the runway at Ramon AB, the jet failed to get out of it and slammed into the runway in front of many of the base's soldiers. *Lt. Col. Eliel* was killed in the accident.

July 1986 - *Lt. Col. Rani Falk* was appointed as the

▼ Lt. Col. Gideon Livni the first F-16 squadron commander.

▼ Captain Ilan Ramon and Nimrod Gur playing checkers at the squadron club.

new squadron commander after the accident in which *Lt. Col. Yossi Eliel* was killed.

March 1988 - *Lt. Col. Rami Falk* handed over command of the squadron to *Lt. Col. Dedi Rosental*.

May 1990 - *Lt. Col. Dedi Rosental* handed over command of the squadron to *Lt. Col. Rafi Berkovich (Berko)*. Under his command, the first advanced operational training course on F-16 "Netz" was initiated. This was the period of the Gulf War in Iraq and the squadron was ready in case Israel decided or was forced to attack. Twenty-five planes waited armed and ready for long-range missions. During this period, the squadron flew dozens of patrol missions along the eastern border and spent many hours in alert standby.

June 4, 1991 - Operation "Black Sea 25" - Two Netz formations of four jets each, one from 253 Squadron with squadron commander *Rafi Berkovich (Berko)* in the lead, and the second from 140 Squadron attacked terrorist targets in the Sidon area. Each plane was armed with 12 Mk.82 GP bombs. It was the heaviest aerial bombardment since the first Lebanon war in 1982. 96 bombs dropped from eight planes, 30 seconds apart from the first to the last, exploded in a chain for about one minute on two hillsides. It had never been done before!

January 1992 - *Lt. Col. Rafi Berkovich (Berko)* handed over command of the Squadron to *Lt. Col. Rami Tidhar*.

August 1994 - *Lt. Col. Rami Tidhar* left command of the Squadron to *Lt. Col. Nimrod Shefer*.

January 17, 1995 - Two Netz formations from the Negev Squadron were in an air-to-air combat training mission. The training took place over the sea, west of Palmachim Beach. At the completion of an intercept, the tail of F-16A #269, which was no. 2 in the "red" formation, hit the canopy of F-16A #290's cockpit, which was no. 2 in the "blue" formation.

The pilots of the two jets ejected and parachuted into the sea. Captain *Danny Everest's* parachute detached, he fell into the sea and was killed. A helicopter on standby at the Palmachim AB with crews of the 669 Rescue Unit was launched and recovered the body of the pilot of Netz #269.

July 1995 - *Lt. Col. Nimrod Shefer* handed over command of the squadron to *Lt. Col. Avi Ya'akobi (Jek)*.

During this period the squadron was trained in the ability to use General-Purpose (GP) bombs as laser-guided bombs (IAF code name "Zar'it") with laser designator pods (IAF code name "Mehuspas") to reach a CEP (Circular Error Probable) level of 15 meters. This capability resulted in a multitude of combat activities in Lebanon for the squadron.

The squadron had improved its night attack capabilities starting to use the "Journal", a night vision system consisting of binoculars attached to the front of the helmet in front of the pilot's eyes and a battery pack mounted on the back of the helmet.

August 3, 1995 - Operation "Iron Eagle" - First operational use of smart bombs. Two smart bombs were dropped on terrorists' position at Jabal Tzafi in Lebanon.

April 11-27, 1996 - Operation "Grapes of Wrath". In the early morning of April 11, Israeli aircraft and artillery began an intensive bombardment of southern Lebanon as well as targets in the Beirut area and in the Bekaa Valley. The declared objective of these attacks was to put pressure on the Government of Lebanon so that it would curb the activities of Hezbollah. Israel conducted air raids on targets which included Katyusha launchers, Hezbollah installations and personnel, as well as vehicles and civilian infrastructures, some of which Israel said were being used for military purposes. During the operation, the squadron flew 80 attack raids. During 1996, following the take-over of F-16B fighters from U.S. Air Force surplus, the "Netz" squadrons began employing navigators. The contribution of the navigators was mainly focused on night missions and formation leading.

On April 14, the squadron (in collaboration with 140 Squadron) deployed to Akinci Turkish Air Force Base near the capital Ankara, as part of the "Selin" exercise. Six F-16s on a non-stop flight, including in-flight refueling, set out for a week-long deployment.

July 1997 - *Lt. Col. Avi Ya'akobi (Jek)* passed command of the Squadron on to *Lt. Col. Shay Prigat*.

Winter 1997 - In Winter 1997 the squadron hosted three Polish MiG-29s for three weeks in an operation named "Sukkah at Night". These MiG-29s were flown by IAF Flight Test Center pilots, and pilots from all the IAF squadrons had the opportunity to train against the MiGs in combat training sorties. After the operation, the MiG-29s were flown back to Poland.

September 7, 1997 - During a training flight in the Negev, a bird hit F-16B #991, flying at a speed of 540 knots and at an altitude of 400 feet. As a result of the impact, a hole about 10 cm in diameter was punched in the canopy and Plexiglas fragments, pieces of the bird and a strong air-stream entered the cockpit. The instructor pilot in the back seat initiated ejection of himself and the new young pilot in the front seat. The crew landed on a steep, rocky slope on the Ramon Crater wall, sustained further injuries and was eventually rescued by helicopter and flown to Soroka Hospital.

May 1999 - *Lt. Col. Shay Prigat* left command of the squadron to *Lt. Col. Alon Ofek*.

May 2001 - *Lt. Col. Alon Ofek* handed over command of the squadron to *Lt. Col. Erez Tzelnik*, who was the last squadron commander in the "Netz" era which ended in March 2003.

May 31, 2003 - The closing ceremony of the squadron and the transfer of its aircraft to the Nevatim AB,

PHOTO GALLERY

Set of frames from *Uri Gil* sight camera downing the MiG-21 using his canon at 8.2g.

▼ Before the hit, the exhaust plumes of two missiles strips that the Syrian pilot launched on Sharabi's F-16 are clearly visible.

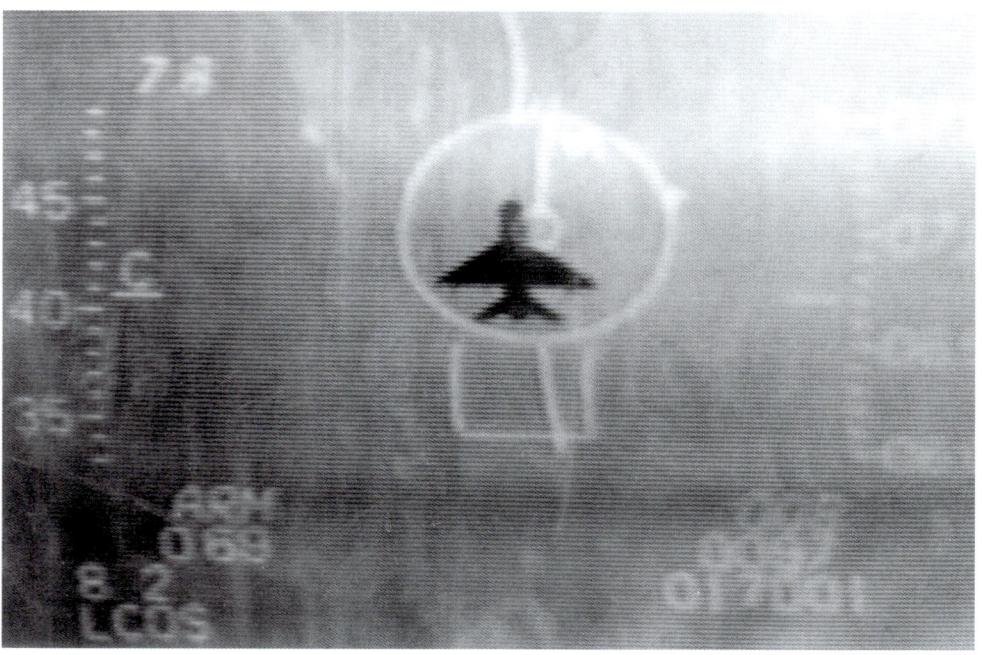

▼ Before the hit.

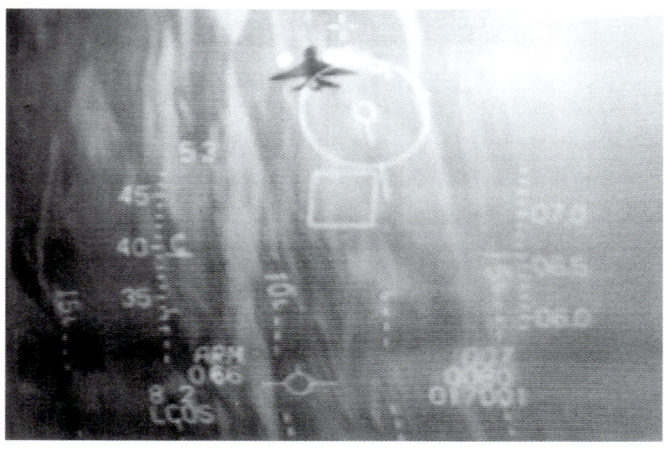

▼ The hit.

▼ After the hit.

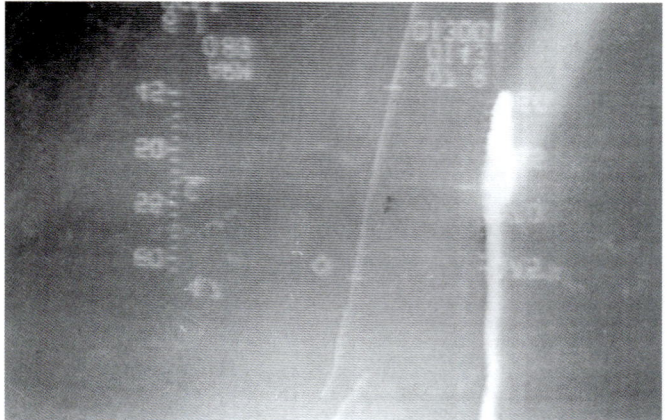
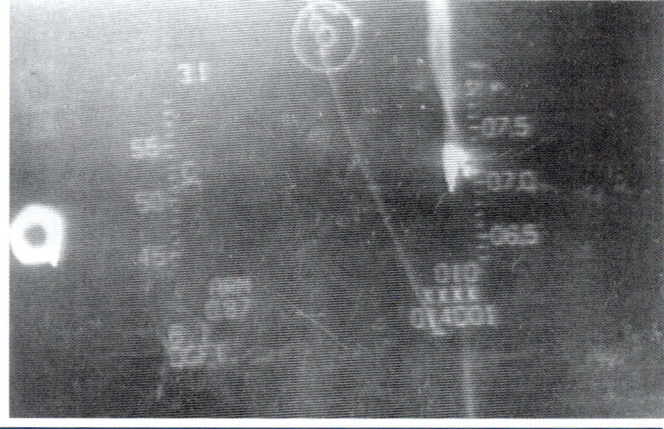

The squadron's deputy commander, major Amos Yadlin, in the squadron's operations room. ▶

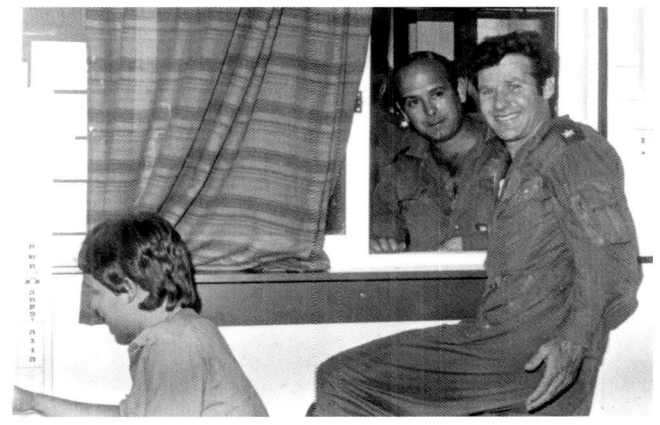

▼ A day of aircraft cleaning in the squadron, ground crew and air crew got together and formed the squadron number on the apron.

▼ Ground crews are washing F-16 #131 that was transferred to them from 117 Squadron.

◄ Work meeting. From right to left: Squadron Commander Lt. Col. Gideon Livni and his deputies captain Ofer Safra and major Amos Yadlin.
(Photo: Ofer Safra collection).

◄ July 1983 - Capt. Amir Eshel (IAF Commander 2012-2017) drenched, after the traditional hose down upon his return from his first solo flight on F-16 along with the protective aircraft shelter ground crew.
(Photo: Pini
Elmakiyes collection).

◄ Major Ofer Safra Getting ready to fly wearing a G-suit during the Gulf War.
(Photo: Ofer Safra collection).

Air crew of 253 Squadron some time between 1989 and 1991. ▶
(Photo: Ofer Safra collection).

▲▼ F-16A #111 is seen here before takeoff for a training flight with the Polish MiG-29 during operation "Sukkah at Night, in 1997. On that occasion, three MiG-29s were deployed to Ramon AB to let the Israeli pilot assess that fighter type. (Photos: Pini Elmakiyes collection).

◀ F-16A #116 flying over the Negev.

▼ F-16A #107 and #126 is seen here in close formation flying over the Negev.
Photo: Dubi Tal - Albatross Aerial Photography

▼ F-16A #129 ready for its next training mission.

▲ Close up on the tail section of the Polish MiG-29 with F-16A #135 in the back. (Photo: Pini Elmakiyes collection).

◀ Close up on the right tail section of F-16A #129.

◀ F-16A #135 awaiting maintenance outside its protective shelter.

▼ F-16A #135 "Studying the enemy" the MiG-29.

◀ F-16A #257 performing a low level fly-by in an air show on Israeli Independence Day. (Photo: Dan Hadani Collection National Library of Israel)

▲ F-16A #260 while refueling in-flight on the way to deployment in Turkey - Spring 1997.
(Photo: Pini Elmakiyes collection).

◀ F-16A #261 carrying four Mk.84 bombs and a pair of AIM-9L air-to-air missiles.

▲ F-16A #264 carrying four AIM-9 air-to-air training rounds. (Photo: Major E.G. collection).

▲ F-16A #267 while refueling in the air on the way to deployment in Turkey - Spring 1997. (Photo: Major E.G. collection).

▲ F-16A #267 armed with Mk.82 bombs and AIM-9L missiles, during Operation Accountability in June 1993. (Photo: Pini Elmakiyes collection).

▲ F-16A #269 - exhibit in static air show at Ben Gurion airport (May 1987).

▲ F-16A #273 armed with M117 General-Purpose (GP) bombs and AIM-9L missiles, As part of an operational deployment at the Ramat David AB in 2001 (note the Pave Penny laser designating pod "Mehuspas").
(Photo: Major E.G. collection).

◄ F-16A #274 at the hot refueling point, awaiting take off order,

F-16A #274 in a 1998 photo with the large star of David Air Force insignia applied on both sides of the vertical fin and on the top and bottom wings David (the large star of David was used to mark adversary aircraft during air combat training). The IAF 50th anniversary big logo applied only on the left side of the fin. ▶

F-16A #276 still with its original factory tail number. ▶

F-16A #277 armed with AIM-9L missiles takes-off from Ramon AB. ▶

◄ F-16A #285 armed with four Mk.84 2,000 lb GP bombs and AIM-9L missiles. Note the AN/ALQ-119 ECM pod at the centreline station. At the hot refueling point, awaiting take off order, during Operation Accountability in June 1993. (Photo: Major E.G. collection).

◄ F-16A #287 still with its original factory tail number.

◄ F-16A #287 after the "Falcon F" program (page 8) with the new tail number style.

F-16A #287 accident after skidding prepare during taxing and hitting the protective shelter wall. ▶

▼ *F-16A #107 - exhibit in static air show at Hatzerim AB -1997. Note that the armament on display included the Python 4 missile, which was never integrated into the Netz.*

Major General Eitan Ben-Eliyanu (IAF Commander 1996-2000) visited the Negev Squadron and joined a training mission sortie in the back seat of an F-16B. (Photo: Major E.G. collection). ▶

▲ *F-16B #017 awaiting maintenance outside its protective shelter.* ▼

▼ *Peace Marble IV - F-16B #979 still in its original USAF camouflage colors.*

▲ F-16B serial 979 from 253 Squadron was one of the aircraft which deployed to Italy in September 1999 in operation "The Beautiful Life". Note the three-tanks long range configuration, including two 600 gal. under wing tanks. (Photo: Riccardo Niccoli archive).

▼ April 1996 - The ground crew (253 and 140 Squadrons) with the 253 Squadron Commander, Lt. Gen. Avi Yaakobi, in the first deployment to Turkey. (Photo: Pini Elmakiyes collection).

A series of beautiful photographs taken in honor of the squadron 25th anniversary.
(Photos: Dubi Tal - Albatross Aerial Photography)

March 31, 2003 - The Netz era in the Negev Squadron is over. The squadron F-16s are transferred to Nevatim AB to open the new 116 - Netz squadron. (Photos: Pini Elmakiyes collection).

IAF Commander Maj. Gen. Danny Halutz (IAF Commander 2000-2004) is also present at the event and transferred F-16B #983. ▶

▲ *Group photo of the ground and air crews of the Negev Squadron. (Photo: Major E.G. collection).*

The Golden Eagle Squadron

The 140 Golden Eagle Squadron was established on July 27, 1959 as an advanced pilot training squadron within the Israeli Air Force Flight School operating North American T-6 "Texan/Harvard" trainer aircraft. In November 1952, the squadron was given an additional task as a light attack squadron during a situation of war. As a matter of fact, the squadron participated in 1956 Operation Kadesh and carried out 24 ground attack raids.

On May 15, 1959, the squadron was deactivated. It was reactivated at Etzion AB on August 1st, 1973, again as an advanced pilot training squadron and assault squadron in wartime, with the landing of the first six A-4E Skyhawks transferred from 116 Squadron operating from Tel Nof AB. The squadron became the largest squadron in the Air Force with 48 A-4 Skyhawks.

In July 1985, the Skyhawk era in the Golden Eagle Squadron ended. In November 1986, a new era began, 140 became the second F-16 fighter squadron based at Ramon AB.

August 3, 1986 - The opening ceremony of the 140 "Netz" Squadron with the participation of the IAF Commander, *Major General Amos Lapidot*, Ramon AB Commander, *Colonel Aviham Sela*, squadron commander *Lt. Col. Yaron Bul*, officers and squadron personnel.

The squadron re-establishment team included the squadron commander *Lt. Col. Yaron Bul*, his deputies *Oron Uriel* and *Boaz Nehorai*, with *Daniel Flexer* as systems officer. They were gradually joined by other pilots, amongst them *Yosef Matzkevitz (Mitz)*, *Ehud Biderman*, *Daniel Maayan* and *Yishai Refaeli*.

The flights were initially performed as part of the 253 Squadron and with 253 Squadron aircraft. Only on November 27, 1986 the 140 Squadron received the first of its F-16s, jets that transferred to Ramon AB from Squadron 117 in Ramat David.

December 7, 1986 - The day the squadron began flying as an independent unit. A formation of four F-16s took off as part of an air combat training exercise in the Mitzpe Ramon area, the formation leader was *Major Boaz Nehorai*, nr. 2 was *Meir Most*, nr. 3 *Amir Ivtzan* and nr. 4 *Daniel Flexer*. *Major Boaz Nehorai* experienced an engine malfunction in F-16A #225, it began to stutter but it seems that *Nehorai* solved the problem and the aircraft returned to function. Given the danger that the fault could recur, *Nehorai* decided to return to base. He arrived at high speed and had to go around and re-enter the pattern. But even this attempt failed, and he made another go-around. During the pattern, the engine stalled again, and he was forced to land immediately. The aircraft touched the runway at high speed, skipped over the arrestor cable and continued at high speed into the safety net. The aircraft ripped through the net and overturned in a ditch on the side of the runway, catching fire. *Nehorai* was killed in this accident and the aircraft was completely destroyed in the fire.

October 5, 1987 - As part of a wing exercise, a formation of four F-16s took off for an air-to-ground mission. *Captain Gil Ivri*, who was number 3 in the formation, flying F-16A #121 hit mount Shazar just four meters below the ridge line. The pilot was killed.

October 10, 1987 - First operational attack mission. Nes-Harim 1 (Miracle Mountain 1) was the name of the operation. A sneak attack on targets of the George Habash organization, Popular Front for the Liberation of Palestine east of the town of Yanta. The squadron commander *Yaron Bul* was the leader of the four F-16A formation, *Daniel Maayan* was Nr. 2, *Oron Oriyol* Nr. 3 (F-16A #129) and *Nir Padan* Nr. 4. The attack was in missile-protected area and when Nr. 3 *Oron Oriyol* broke contact after the bombing, he was hit by an SA-7 shoulder-fired, low-altitude SAM. The "Netz" went into spin and the engine lost power. *Oron* managed to straighten the jet and maintain a speed of 200 knots until landing at Ramat David AB.

October 20, 1987 - *Lt. Col. Yaron Bul* handed over command of the Squadron to *Lt. Col. Gil Dekel*.

January 1988 - The "Netz" Flight Simulator Division transferred from Ramat David AB to Ramon AB and its operation was entrusted to the 140 Squadron.

April 21, 1988 - In the large aerial air-show on the occasion of the 40th Independence Day of the State of Israel, the squadron was chosen to represent the "Netz" era in the Air Force and five F-16s in arrowhead formation flew from Ramon AB to Kiryat Shmona and back.

December 18, 1988 - Loss of F-16B #015. On a stormy and rainy Sunday, at noon, *lieutenant Rami Ben Ephraim* (later Brigadier General and head of the human resources division of the IAF), took off from Ramon AB for a routine air-to-ground training flight. He then was a young pilot in his operational training, and was number two in a quartet formation that flew over the Judea Desert at a speed of 800 km/h. They flew north along the border with Egypt, after 70 kilometers they turned east, reached the border with Jordan, and lowered to 200 feet. Lieutenant Rami suddenly noticed a huge black lump emerging in front the canopy of his F-16. He recalls: *"At first I did not understand what it was, then I realized it was a bird. Only later I found out that it was a rock eagle (a large raptor, whose wingspan can reach up to 2.3 meters). I was sure that the close encounter between us was inevitable. Either I fly into the bird, or the big bird gets into me. At the last second, the collision was averted. The bird slipped on my left towards the wing. I had goose bumps. I had an early feeling that the collision is inevitable. Suddenly I saw second bird in front of the canopy, of exactly the same kind, and I thought that was it... She did not hit the canopy, but just passed below me towards the engine. In the cockpit you hear nothing. You sit protected in a bubble and listen to the silence around you. Suddenly there was a huge explosion. The engine let out a scream of death, screeched, and started to wind down. It was clear to me that the engine was gone. At that moment I knew that I had to do two things quickly: climb - pull the stick to get away from the ground, and report to the formation. In a moment it became clear to me that I could not do that either. The bird that penetrated the engine broke a fan blade. As a result of the centrifugal force, the blade detached and passed through the engine casing into the fuselage - where it damaged the fuel and oil tanks. As a result, an acute nose up attitude was caused. The aircraft made a loop in the air and then went into a spin. In such a situation one should eject. I had to activate the*

▲ 140 Squadron establishment team. Standing from right to left: Daniel Ma'ayan, Alon Ofek, Dan Ronen, Boaz Nehorai, Yaron Bul (140 Squadron Commander), Rani Falk (253 Squadron Commander), Yosi Lapid, Eyal Eshel and Daniel Flekster.
Kneeling from right to left: Ishai Refaeli, Yosef Matzkevits, Amir Eshel, Yoav Yavne, Oron Oriel and Ehud Biderman. Seating in front: Eitan Bernstein.

▼ Remains of F-16A #121.

▼ Remains of F-16B #015. (Photo: Major E.G. collection).

ejection seat by pulling the handle between my legs, an operation that was supposed to be simple, but I could not get it done. I had a blackout. When this happens, you fly with your eyes open but see nothing. The acute nose up and over means pulling 17gs, i.e., the earth's gravity relative to the aircraft and what is inside increases 17 times. The muscles are unable to cope with the condition, blood does not reach the eyes and brain and you may lose consciousness. I tried desperately to get my hands on the ejection handle and pull, but I could not find it because I lost my sight. I could not raise my hands because of the load on them, which created a feeling of complete paralysis. A terrible feeling of lack of control over your body. I was on the verge of fainting, but I knew I could not lose consciousness. At one point I was able to recover, I noticed the handle of the ejection seat, but I could not drag my hand to it. Apparently in the face of death, the body does everything to survive. It is not clear where I was able to mobilize the forces from, suddenly I was able to drag one hand to the ejection handle and I pulled." Lieutenant Rami ejected from the aircraft when it was upside down and diving, at 30 degrees below the horizon and with 2 seconds left to impact. Luckily, the ejection seat rocket lifted him to a safe height. He separated from the seat and fell on a cliff of Nahal Arugot, while the parachute dragged him along face down. Rami was dragged on the rocky ground for hundreds of meters due to the strong wind. He was wounded all over his body, his visor was broken, and he lost his watch. He was able to detach one buckle with the right hand as his left hand was paralyzed. He was dragged another 200 meters until he managed to lower the parachute dome. After 30 minutes, a CH-53 Yasur helicopter arrived with a rescue team that took him to the hospital. Luckily, the injuries were superficial. F-16B #015 however was lost.

August 20, 1989 - Lt. Col. Gil Dekel passed command of the squadron to Lt. Col. Ido Nehushtan.

August 1, 1990 - The first Advanced Operational Training Course (AOTC) with five young pilots started as a new separate unit within the squadron.

During the period of the Gulf War in Iraq, the squadron was ready in case Israel decided or had to attack. The squadron's fighters waited for hours in standby, armed and ready for any intercept scramble.

In addition, the squadron also conducted air patrols and until February 28, the date on which the ceasefire was announced, it flew about 60 sorties.

June 4, 1991 - Operation "Black Sea 25" - The 140 Squadron formation was led by Ofer Green, with him Lihu Ha'Choen, Mark Bregman and Alon Shavit. (Details of the operation in the chapter on the Negev Squadron).

July 1st, 1991 - Lt. Col. Ido Nehushtan handed over command of the squadron to Lt. Col. Adi Gur-Lavi (Gorla).

July 25-31, 1993 - Operation "Accountability" - The squadron flew 18 assault missions during the operation when in one of the missions a section of the squadron fired two AGM-65 Maverick air-to-surface missiles (named "Penguin" in the IDF/AF) at the target. Thus, they opened the era of employing the Penguin with the F-16 fighters in the IAF.

August 10, 1993 - Lt. Col. Adi Gur-Lavi (Gorla) passed on command of the squadron to Lt. Col. Amir Chodorov.

January 1995 - At the end of 1994, the IAF received 50 F-16A/Bs from the U.S. Air Force surplus as a gift, which entered service with the 144 Squadron at Hatzor AB. Some of the F-16B planes also arrived at Ramon AB and enabled the opening of the first experimental OTC (Operational Training Course) in F-16 jets, normally performed on the A-4 Skyhawk.

Amir Chodorov recalls: *"Everything was a coincidence. Air Force commander Herzl Bodinger asked me during one of his visits to the squadron whether I considered it possible to perform the operational training course with the F-16 jets. I replied that anything could be done but the subject should be studied. The F-16 is a nimble and fast plane and not as forgiving in case of mistakes like the A-4 Skyhawk".*

The subject was not raised anymore. A few weeks later Bodinger invited me to his office and told me: 'I am invited to visit the Dutch and British Air Force and would like you to join me'.

Of course, the commander of the Air Force cannot be refused a request.

When we visited the Dutch Air Force it turned out that they were doing the conversion course on F-16s... they had nothing else. And they were sure we also do it on the F-16s.

"I tried to learn something from them about their way of working but I could not..., they were not very open with us.

After we returned, I was informed that the commander of the Dutch Air Force would be visiting us, the visit would be to my squadron and I should immediately devise a conversion course plan and select five candidates who would start the course. And that is exactly what I did. I prepared a training syllabus and together with the flight school commander we selected five pilots and I opened the first course.

Despite the difficulties the course passed successfully and immediately after that, in July 1995 we opened another course in which six pilots participated. And that was exactly when the Dutch Air Force commander and his entourage came to visit.

At the end of the second course, it was decided to discontinue these courses in the squadron because there were still A-4 Skyhawks used for this purpose and it was a little too early. Instead, it was decided to carry out the advanced training course on F-16s. For about four years the Air Force worked on the issue well, made its conclusions, and then when the

Skyhawk planes were taken out of service, the course was regularly transferred to the F-16s".

During this period, the squadron was trained in the ability to use General-Purpose (GP) bombs with laser-guiding packages (IAF code name "Zar'it") with the Pave Penny laser designation pod ('Mehuspas').

August 1995 - *Lt. Col. Amir Chodorov* handed over command of the squadron to *Lt. Col. Ofer Green*.

April 11-27, 1996 - Operation "Grapes of Wrath".
The main activity of the squadron in Operation Grapes of Wrath were attacks with laser-guided bombs called "Mehuspas sorties" in the IAF.
Of the 87 operational sorties during the operation, 82 were "Mehuspas sorties".
At the same time, on April 16, some squadron pilots were flown to Akinci Turkish AFB in a C-130 Hercules to replace pilots from the 253 Squadron deployed there as part of the "Selin" exercise.

May 20, 1997 - An exciting and unique farewell ceremony took place in the squadron for *Colonel Giyora Even (Epstein)*, fighter ace credited with 17 victories, making him the ace of aces of supersonic fighter jets and of the Israeli Air Force. *Giyora Even* received the Medal of Distinguished Service, one of Israel's highest military honors. In 1988, *Giyora Even* completed a conversion course on the F-16 as a reserve pilot and continued to serve in 140 Squadron as a reserve until his 59th birthday, the day he retired. An aerial display that included 24 F-16s was organized. F-16A #254 was decorated with 17 kill marks and its tail number changed to 140, it was this Netz that *Giyora Even* piloted during the display.

August 25, 1997 - *Lt. Col. Ofer Green* leaves command of the squadron to *Lt. Col. Erez Sason*.

March 1999 - First deployment in the history of the Israeli Air Force confronting US Navy fighter jets. Twelve F-16s from the squadron were deployed at Ovda AB, joined by three F-14s and nine F-18s from the aircraft carrier USS Enterprise, which was anchored off the shores of Ashdod. This week-long deployment, in which individual and combined air battles were fought, ended in an impressive Israeli victory of 200 kills for 140 pilots and 11 victories for the US Navy pilots.

May 12, 1999 - *Lt. Col. Erez Sason* handed over command of the Squadron to *Lt. Col. Rami Ben-Efraim*.
During this period OTC "Netz" goes from experiment to routine, and the squadron actually returned to its initial mission - OTU (Operational Training Unit) simultaneously with its tasking as an operational squadron.

September 1999 - First deployment to Italy with operation code name "The beautiful life". A pair of F-16Bs from the squadron, a pair of F-16Bs from 253 Squadron and a pair of F-15Is from 69 Squadron deployed to Amendola AB for one week of navigation training missions.

February 2, 2000 - As part of a routine aerial photography training within the advanced training course, *Gal Vegmeister* took off as number 2 in the formation. While getting ready for aerial photography at an altitude of about 14,000 feet, an explosion was heard in the engine and it shut down. After two unsuccessful starting attempts, a fire broke out and at an altitude of about 3,000 feet, he decided to abandon the aircraft. *Gal* ejected successfully and F-16A #237 crashed. The investigation of the accident revealed that the a blade of the engine fan broke, and the engine disintegrated in the air.

August 1, 2000 - Back from air defense training mission, *Yaniv Dalal*, flying F-16A #248, hears an explosion in the engine, the aircraft stops for a moment mid air and then begins to dive sharply. *Dalal* immediately ejected at a high altitude and landed safely. The aircraft crashed, and the accident investigation reveals that an entire fan disintegrated and totally destroyed the engine.

May 2001 - *Lt. Col. Rami Ben-Efraim* hands over command of the squadron to *Lt. Col. Ronen Simchi*.

May 2003 - *Lt. Col. Ronen Simchi* left command of the squadron in the hands of *Lt. Col. Izhar Merchavi*.

June 2003 - Following the Air Force's acquisition of F-16I "Sufa" fighters, it was decided to operate them from Ramon AB. 140 squadron was transferred to Nevatim AB.

November 2003 - Five IAF F-16s fighter jets are deployed to Turkey, including one from 140 Squadron (#010).

May 2005 - *Lt. Col. Izhar Merchavi* handed over command of the squadron to *Lt. Col. Avshalom Amosi*.

May 2007 - *Lt. Col. Avshalom Amosi* transferred command of the Squadron to *Lt. Col. Gay Waisberg*.

May 2009 - *Lt. Col. Gay Waisberg* handed over command of the squadron to *Lt. Col. Or Landver*.

September 13, 2009 - Some F-16s of the "Golden Eagle" Squadron took off with advanced training course pilots from pilot course 158 for air-to-air combat training. Lieutenant *Assaf Ramon*, the son of the first Israeli astronaut, the late *Col. Ilan Ramon*, flying F-16A #140 (236) was nr. 2 in the formation. Above Mount Hebron, at an altitude of 18,000 feet., the formation leader noticed *Assaf* diving straight for the ground. He called *Assaf* but no response was heard. The aircraft dived for what seemed eternal seconds until it hit the ground near the settlement of Bnei Hever. *Assaf Ramon* was killed on the spot. He was promoted to the rank of captain posthumously.

May 2009 - *Lt. Col. Or Landver* leaves command of the squadron in the hands of *Lt. Col. Omer Tishler*.

August 5, 2013 - In the summer of 2013, it was decided to deactivate 140 Squadron as a "Netz" unit. In an impressive ceremony, the squadron closed, and its aircraft were transferred to 116 Squadron.

PHOTO GALLERY

F-16A #100 looks naked here without any armament and the missile launchers. ▶

▼ An amazing photo of F16A #109 as the pilot inside the cockpit awaits approval to taxi out for take-off. (Photo: Major E.G. collection).

▼ Photo from January 3, 2009. F-16A #109 during operation "Cast Lead" loaded with four M117 General-Purpose (GP) bombs. (Photo: Major E.G. collection).

▲ ▼ F16A #105 in a late 1998 photo with the large star of David Air Force insignia applied on both sides of the vertical fin and on the top and bottom wings. The IAF 50th anniversary patch applied only on the left side of the fin.

▼ *A series of three photographs taken on October 29, 2012, of F-16A #109 on its way out for a routine training flight. (Photos: Sariel Stiller).*

▲ Photo from January 3, 2009. F-16A #113 during operation "Cast Lead" loaded with four M117 General-Purpose (GP) bombs. (Photo: Major E.G. collection).

▼ F-16A #113 on its way to a training sortie.

▲ F-16A #113 after take-off on its way to a training sortie. (Photo: Hezi Shmueli).

▼ Ground crews prepare the pilot in F-16A #113 for departure on a training sortie. (Photo: Dolev Gottdiener).

◀ F-16A #114 photo from 1998 at the hot refueling point, waiting take off order.

▼ F-16A #114 in a static ground exhibition loaded with two inert Mk.84 bombs. (Photo: Moni Shafir).

▼ F-16A #118 taxiing into its protective shelter, returns from a training sortie - Ramon AB.

▲ "Netz" #118 in preparation for operational attack mission, loaded with two Mk.84 bombs - Nevatim AB.

▲ "Netz" #118 taxiing for take-off - Nevatim AB.

▲ F-16A #118 and F-16B #763 retired from service and stored at Ovda AB. (Photo: Major E.G. collection).

◀ F-16A #126 waiting for maintenance - Ramon AB 1987.

F-16A #126 during take-off, Nevatim AB 2013.
(Photo: Moni Shafir). ▼ ▶

▲ F-16A #219 parked in the squadron area outside its protective shelter.

▲ F-16A #219 taxiing for take-off. (Photo: Moni Shafir).

▲ F-16A #219 just after take-off. (Photo: Moni Shafir).

◀ F-16A #228 after taking off. Note Elta EL/L-8212 ECM pod on the central rack.
(Photo: Hezi Shmueli).

Photo: Moni Shafir.

◀ A quartet of F-16As in first daylight, ready and waiting for their pilots.
(Photo: Major E.G. collection).

▲ F-16A #230 in first daylight, ready and waiting for its pilot. (Photo: Major E.G. collection).

▲ Photo from January 3, 2009. F-16B #006 during operation "Cast Lead" loaded with four M117 General-Purpose (GP) bombs. (Photo: Major E.G. collection).

◀ F-16A #232 approaching for landing.
(Photo: Major E.G. collection).

▲ "Netz" #233 taxiing for take-off. (Photo: Moni Shafir).

◀ F-16A #236 after its tail number was changed to 140.

F-16A #243 in full afterburner. (Note the Pave Penny laser designating pod "Mehuspas") (Photo: Major E.G. collection). ▶

F-16A #243 while refueling in the air from a Re'em B707 tanker on the way to operation "Beautiful Life" in Italy - 1999. ▶
And in close formation fl in the skies over Italy. (Photo: Major E.G. collection). ▶

◄ F-16A #243 with a friendly pilot.
Nevatim AB - May 17, 2015
(Photo: Riccardo Niccoli).

▲ "Netz" #243 and #252 taxiing for take-off. Nevatim AB - August 5, 2013. (Photo: Moni Shafir).

◄ F-16A #243 during maintenance inside its Hardened Shelter (HAS) - Nevatim AB - May 9, 2013.
(Photo: Dolev Gottdiener).

F-16A #246 In preparation for loading armament for an attack configuration. ▶
(Photo: Dolev Gottdiener).

Photo: Moni Shafir.

F-16A #246 close look. ▶
(Photo: Dolev Gottdiener).

▲ "Netz" #252 taxiing for take-off. Nevatim AB - March 8, 2013. (Photo: Moni Shafir).

▲ "Netz" #254 taxiing for take-off for an operational attack mission, armed with four live AIM-9L missiles and two Mk.83 1,000 lb GP bombs. Nevatim AB - July 22, 2013. (Photo: Moni Shafir).

◄ F-16A #254 with an AN/ALQ-119 ECM pod on the way out for a training sortie.

F-16A #255 awaiting its pilot. (Photo: Dolev Gottdiener). ▶

F-16A #258 exhibit in static display (1998). ▶

▼ *F-16A #258 in a beautiful take-off shot. (Photo: Lior Kestner).*

▲ "Netz" #258 and #126 exhibit during an open day. Nevatim AB - March 8, 2013. (Photo: Moni Shafir).

◀ F-16A #261 in a static exhibition during an open day at Tel Nof AB. Note the bomb rack IMI VER-4 for four bombs, that was never introducted into service.

▲ "Netz" #272 applies full power for take-off. (Photo: Moni Shafir).

▲ "Netz" #272 approaching for landing with its wing men escorting him. Nevatim AB - March 8, 2013.
(Photo: Moni Shafir).

F-16A #274 on the squadron flight line. ▶

"Netz 2" #733 previously belonged to 144 Squadron, on the day the squadron moved to the Nevatim AB.
(Photo: Pini Elmakiyes collection). ▶

▲ "Netz 2" #745 previously belonged to 144 Squadron in the IAI scrapyard. (Photo: Moni Shafir).

▲ "Netz 2" #747 previously belonged to 144 Squadron on the day of the closure of Squadron 144 and the transfer of its aircraft to the Nevatim AB October 10, 2005. (Photo: Sariel Stiller).

"Netz 2" #751 previously belonged to 144 Squadron. ▶

November 15-25, 2010 - Italian and Israeli jets deployed to Decimomannu air base in the island of Sardinia for the bi-national exercise VEGA 2010. Three F-16B from the Golden Eagle Squadron participate in the exercise (#001, 006 and 981).

Photo: Major E.G. collection

Photo: Major E.G. collection

Photo: Major E.G. collection

Photo: Major E.G. collection

Photo: Giovani Colla

101

Photo: Massimo Pieranunzi

Photo: Massimo Pieranunzi

Photo: Giovani Colla

▲ "In the front of the IAF F-16s, the first Italian special colors F-16, MM.7251, was realized by the 23 Gruppo, to celebrate the 90th anniversary of the squadron. (Photo: Major E.G. collection)

▲ "In front of the IAF F-16s, Italian Air Force F-16A MM.7249 from 10 Gruppo, in special colors celebrated the 1,000 hours on the type. (Photo: Major E.G. collection)

Photo: Major E.G. collection

Photo: Major E.G. collection

Photo: Major E.G. collection

Photo: Massimo Pieranunzi

Photo: Massimo Pieranunzi

Photo: Massimo Pieranunzi

Photo: Massimo Pieranunzi

Photo: Massimo Pieranunzi

Photo: Massimo Pieranunzi

Photo: Massimo Pieranunzi

Photo: Massimo Pieranunzi

Photo: Massimo Pieranunzi

Photos: Major E.G. collection

▲ September, 1999 - First deployment to Amendola AB Italy. Operation "The beautiful life". ▲ ▼
(Photos: Major E.G. collection).

▼ "Netz" #010 climbing in full power after take-off. (Photo: Moni Shafir).

▲ F-16B "Netz 2" #981 taxiing for take-off. Nevatim AB - March 8, 2013. (Photo: Moni Shafir).

▲ F-16B "Netz 2" #984 climbing in full power after take-off. (Photo: Major E.G. collection).

▲ F-16B "Netz 2" #990 previously belonging to 144 Squadron in the IAI scrapyard. (Photo: Moni Shafir).

▲ F-16B "Netz 2" #995 in a beautiful take-off shot. Nevatim AB - March 8, 2013. (Photo: Moni Shafir).

▲ F-16B "Netz 2" #995 just after landing Nevatim AB - April 21, 2013. (Photo: Hezi Shmueli).

▲ F-16B "Netz 2" #998 with its ground crew completing last checks before take-off.

May 20, 1997 - Last flight and farewell ceremony from Colonel Giyora Even (Epstein), fighter ace credited with 17 victories, making Epstein the worlds' ace of aces of supersonic fighter jets and of the Israeli Air Force.

Photos: Brigadier General Giyora Even (Epstein) collection.

◀ An exciting photo in which Giyora Even (Epstein) returns his original pilot wings to the then President of Israel Mr. Ezer Weizman who gave them to Giyora at the end of the IAF pilot course back in 1963.

F-16B #010 approaching for aerial refueling from a Turkish Air Force KC-135 tanker during the deployment to Turkey on November 2003.
(Photo: Major E.G. collection). ▶

▲ 140 Squadron Closing Ceremony - August 5, 2013. (Photos: Moni Shafir). ▼

A series of beautiful photographs taken in honor of the squadron anniversary.
(Photos: Dubi Tal - Albatross Aerial Photography)

 # The Phoenix Squadron

On January 10, 1972, an order was issued to establish the new 144 IAI Nesher Squadron. On September 6, 1972, with the landing of the first six IAI Nesher jet fighters at Etzion AB, the squadron was opened as an intercept unit.

During the Yom Kippur War the squadron made about 700 sorties and its pilots shot down 40 enemy aircraft.

On December 1, 1978, the squadron converted from IAI Nesher to the IAI Kfir C-2 jet fighters. The IAI Neshers that had served in the squadron since its inception were sold to Argentina.

Following the peace agreement with Egypt and the closure of the Etzion AB, on January 10, 1982 the squadron was transferred to the new Ovda AB. During the Peace for Galilee war the it deployed to the emergency AB near Megiddo and operated from there. During the operation, the squadron carried out 158 support sorties, interdiction missions, intercept patrols, reconnaissance escorts and SAM battery suppression missions. In May 1983, the squadron converted to the improved IAI Kfir C-7 fighter jets and became the only squadron in the IAF to operate the C-7 model. In July 1988, the squadron was transferred to Hatzor AB, and in June 1994, the Air Force Commander Major General *Herzl Bodinger* decided to decommission the IAI Kfir jets and re-equip the squadron with F-16 fighters. On August 1st, 1994, the squadron reactivated with the landing in Israel of the first six F-16s (named in the IAF "Netz 2") from the U.S. Air Force surplus. Lt. Col. *Elisha Hosman* was appointed the first commander in the squadron F-16 era.

Elisha Hosman recalls: "The "Surplus Netz" squadron was to be formed at Nevatim AB, using the facilities of the 115 Squadron, which was deactivating. This was decided by the commander of the Air Force, Major General Herzl Bodinger. The founding team of the 144 Netz Squadron Hagai Meyer, Ehud Levy and myself held the squadron's basic meeting at a cafe in Be'er Sheva. At some stage, probably due to an economic re-assessment, it was decided to return to Hatzor AB, the home of the deactivated 144 IAI Kfir squadron. A month and a half later, on August 1, 1994, the first fighter jets landed, flown in by American pilots. The 50 aircraft, 35 F-16A single-seaters, and 15 F-16B two-seaters came from two sources. Some of them originated from an active

▲ *F-16A #750 on final for landing - Hatzor AB 1996. (Photo: Maurits Even).*

▼ *F-16A #750 in flight in 1998 still in standard USAF camouflage colors - (Note the IAF 50th anniversary sticker applied on the left side of the fin).*

US National Guards squadron - aircraft that were characterized by a fair level of serviceability and some modifications. The other part were planes that had sat for several months in an "airplane cemetery". After being personally selected by an IAF delegation, they were rendered serviceable and bravely flown across the ocean. The planes came with a mechanical engine control, without warning devices and with outdated radio equipment. After a three-week acceptance inspection, we performed a test flight with them, and with the help of the excellent people of the technical department, the planes were accepted in serviceable status. Very quickly we brought the planes up to the level of the other IAF F-16s (these planes had more service hours, but the airframes had been much less solicited)".

The squadron passed the competency review with

notable success and was declared operational. The first attack was not long in coming and on August 14, 1995, *Hosman* and *Ira* bombed an anti-aircraft gun position in Noima. A month later, *Meyer* and *Raichel* took off for their first operational patrol flight to Beirut. In April 1996 Operation Grapes of Wrath commenced. The squadron made more assault raids than all other single-seater squadrons. 44 assault raids, including an attack on the power station near Beirut and approximately 20 air-to-air raids.

November 27, 1996 - A pair of F-16s set out for an air- to-ground assault training sortie. The formation leader took off followed by F-16A #750 piloted by *Ran Meged*. During takeoff when *Ran* moved the throttle to the after-burner position, an explosion was heard, and the engine failed. The pilot took emergency action, shut down the engine, aborted the take-off and stood hard on the brakes. The jet stopped in the middle of the runway. The fire that had broken out in the engine quickly spread forward, reaching the cockpit. *Ran* raised the canopy and opened the harnesses that tied him to the ejection seat, the parachute and survival pack in the seat. He climbed out of the cockpit, jumped to the ground, and walked away from the burning jet.

Squadron 144 led the development of the "Gondola" Elta EL/M-2060P radar photography system to operate on F-16s. The Elta EL/M-2060P pod, containing a Synthetic Aperture Radar (SAR) reconnaissance system, able to provide high-resolution radar images by day or night. The pod had the same shape of a F-16 ventral auxiliary tank.

November 2003 - Five IAF F-16s fighters are deployed to Turkey including two from the squadron (#989 and #993).

Squadron commanders after *Hosman* were:

July 1996 - September 1998 - Lieutenant Colonel *Arik Weissman*.

September 1998 - July 2000 - Lieutenant Colonel *Ehud Biderman*.

July 2000 - August 2002 - Lieutenant Colonel *Alon Shlomi*.

August 2002 - August 2004 - Lieutenant Colonel *Kobi Regev*.

August 2004 - October 2005 - Lieutenant Colonel *Dan Tortan*.

On October 10, 2005, the squadron was deactivated. Some of its aircraft were divided among the F-16 squadrons and the rest were transferred for storage at Ovda AB.

"Gondola" Elta EL/M-2060P. (Photo: Sariel Stiller).

"Peace Marble IV" F-16B

IDF/AF S.N.	C.N.	BLOCK	USAF S.N.	DELIVERY DATE
974	M11-01	5	78-0086	9.9.1994
976	M11-02	5	78-0095	9.9.1994
979	M11-03	5	78-0106	23.9.1994
983	M11-04	5	78-0108	9.8.1994
984	M11-05	5	78-0109	24.9.1994
986	M11-06	5	78-0111	29.9.1994
989	M11-07	5	78-0114	29.9.1994
981	M11-08	5	78-0115	24.9.1994
991	M11-09	5	79-0410	4.11.1994
993	M11-10	5	79-0423	9.8.1994
995	M11-11	5	79-0424	26.7.1994
996	M11-12	5	79-0425	29.9.1994
990	M11-13	5	80-0624	18.11.1994
998	M11-14	5	80-0632	29.9.1994

"Peace Marble IV" F-16A

IDF/AF S.N.	C.N.	BLOCK	USAF S.N.	DELIVERY DATE
776	M10-01	5	78-0012	9.9.1994
702	M10-02	5	78-0014	1.9.1994
705	M10-03	5	78-0018	9.8.1994
708	M10-04	5	79-0288	18.11.1994
709	M10-05	5	79-0289	23.9.1994
712	M10-06	5	79-0291	18.11.1994
714	M10-07	5	79-0292	18.11.1994
721	M10-08	5	79-0293	4.11.1994
724	M10-09	5	79-0295	4.11.1994
726	M10-10	5	79-0297	4.11.1994
728	M10-11	5	79-0299	4.11.1994
729	M10-12	5	79-0302	18.11.1994
731	M10-13	5	79-0304	4.11.1994
733	M10-14	5	79-0305	9.8.1994
735	M10-15	5	79-0319	4.11.1994
737	M10-16	5	79-0320	4.11.1994
738	M10-17	5	79-0321	4.11.1994
740	M10-18	5	79-0325	4.11.1994
742	M10-19	5	79-0328	1.9.1994
745	M10-20	5	79-0333	9.9.1994
747	M10-21	5	79-0339	1.9.1994
750	M10-22	5	79-0347	26.7.1994
791	M10-23	5	79-0356	1.9.1994
751	M10-24	5	79-0358	29.9.1994
752	M10-25	5	79-0361	26.7.1994
754	M10-26	5	79-0369	9.8.1994
755	M10-27	5	79-0377	26.7.1994
757	M10-28	5	80-0491	1.9.1994
759	M10-29	5	80-0501	26.7.1994
760	M10-30	5	80-0502	23.9.1994
763	M10-31	5	80-0503	9.8.1994
765	M10-32	5	80-0514	23.9.1994
766	M10-33	5	80-0516	1.9.1994
769	M10-34	5	80-0517	26.7.1994
771	M10-35	5	80-0532	4.11.1994
777	M10-36	5	80-0534	29.9.1994

PHOTO GALLERY

▲ F-16A #702 with an AN/ALQ-119 ECM pod after take-off (Ovda AB August 8, 2005). (Photo: Major E.G. collection).

▼ F-16A #705 used for firefighting exercises at Nevatim AB. (Photo: IAF Magazine).

▲ F-16A #709 waiting for his turn to take off on the day the squadron closed (Hatzor AB October 10, 2005). (Photo: Sariel Stiller).

▲ F-16A #712 taken while landing, with the first tail design.

◄ F-16A #712 parked in the squadron area with the new tail markings design, introduced in summer 2001.

▲ F-16A #714 with an AN/ALQ-119 ECM pod during take-off (Ovda AB August 8, 2005). (Photo: Major E.G. collection).

▲ F-16A #714 Hatzor AB October 10, 2005. (Photo: Sariel Stiller).

▲ F-16A #721 Hatzor AB October 10, 2005. (Photo: Sariel Stiller).

▲ F-16A #724 now in the IAI scrapyard. (Photo: David Weinrich)

▲ F-16A #726 Ovda AB August 8, 2005. (Photo: Major E.G. collection).▼

▲ *F-16A #726 now in the IAI scrapyard. (Photo: Moni Shafir).*

▲ *F-16A #731 Hatzor AB October 10, 2005. (Photo: Sariel Stiller).*

▲ *F-16A #733 Hatzor AB October 10, 2005. (Photo: Sariel Stiller).*

▲ F-16A #735 after it entered service in its original USAF camouflage colors.

▲ F-16A #735 Ovda AB August 8, 2005. (Photo: Major E.G. collection).

▲ F-16A #735 Hatzor AB October 10, 2005. (Photo: Sariel Stiller).

F-16A #738 in its original USAF camouflage colors with F-16A #755 already painted in the standard IAF camouflage colors. Note the 50th anniversary sticker applied on the fin. ▶

▲ *F-16A #738 Ovda AB August 8, 2005. (Photo: Major E.G. collection).* ▼

▲ F-16A #740 Hatzor AB October 10, 2005. (Photo: Sariel Stiller).

▲ F-16A #742 now in the IAI scrapyard. (Photo: David Weinrich)

▲ F-16A #752 Hatzor AB October 10, 2005. (Photo: Sariel Stiller).

▲ F-16A #754 Ovda AB August 8, 2005. (Photo: Major E.G. collection).

▲ F-16A #754 Hatzor AB October 10, 2005. (Photo: Sariel Stiller).

▲ F-16A #755 Hatzor AB in its original USAF camouflage colors.

▲ F-16A #755 Hatzor AB October 10, 2005. (Photo: Sariel Stiller).

▲ F-16A #742 Ovda AB August 8, 2005. (Photo: Major E.G. collection).

▲ F-16A #757 Hatzor AB October 10, 2005. (Photo: Sariel Stiller).

▲ *F-16A #759 or what is left of it after it was cannibalized for parts.*

▲ *F-16A #760 on static display during one of the IAF open days.*

▲ *F-16A #763 Hatzor AB October 10, 2005. (Photo: Sariel Stiller).*

▲ F-16A #765 at Ovda AB scrapyard. (Photo: Major E.G. collection).

▲ F-16A #769 after take-off for a training sortie.

▲ F-16A #769 Hatzor AB October 10, 2005. (Photo: Sariel Stiller).

▲ F-16A #771 Hatzor AB - in its original USAF camouflage colors.

▲ F-16A #777 Ovda AB August 8, 2005. (Photo: Major E.G. collection).

▲ F-16A #777 Hatzor AB October 10, 2005. (Photo: Sariel Stiller).

▲ *F-16A #791 at the Haifa Technical School AB - May 3, 2006. (Photo: Biaf - Aerospace Magazine).* ▼

▼ *F-16A #757 Hatzor AB October 10, 2005. (Photo: Sariel Stiller).*

▲ F-16B #976 Hatzor AB October 10, 2005. (Photo: Sariel Stiller). ▼

▼ F-16B #981 Ovda AB August 8, 2005. (Photo: Major E.G. collection).

▲ F-16B #976 Hatzor AB October 10, 2005. (Photo: Sariel Stiller).

◀ F-16B #993 air refueled by a IAI modified Boeing 707 tanker during the deployment to Turkey in November 2003.
(Photo: Major E.G. collection).

▼ F-16B #989 In 1998 or 1999 prepared for a night flight sortie in an Weapons of mass destruction (WMD) scenario exercise.

▲ F-16B #989 was one of the two Netz which were sent to Izmir air base, Turkey, in June 2001 to take part in the air show for the 90th Anniversary of the Turkish Air Force. Note the two 500 lb Griffin Laser Guided Bombs for training. (Photos: Riccardo Niccoli). ▼ ▼

▼ F-16B #989 Hatzor AB October 10, 2005. (Photo: Sariel Stiller).

▼ F-16B #993 - during the deployment to Turkey - November 2003. (Photos: Major E.G. collection).

▲ ▼ *F-16B #993 - photos from the deployment to Turkey - November 2003. (Photos: Major E.G. collection).*

▼ ▼ *F-16B #996 - was the other aircraft from 144 Squadron which took part to the air show at Izmir for the Turkish Air Force 90th Anniversary. Both the aircraft were equipped with the 600 gallons under wing tanks.*

Photo: Major E.G. collection

Photo: Riccardo Niccoli

▼ *Close-up of an AN/AAS-35 Pave Penny targeting pod, which was used by the Netz for the use with Precision Guided Munitions (PGM). Following the First Gulf War of 1991, also the Israeli F-16A/B fleet was cleared to operate with LGB bombs. (Photo: Riccardo Niccoli).*

▲ *A 500 lb Griffin Laser Guided Bomb attached under the wing of a two-seater Netz. The blue color of the bomb body means it is an inert bomb for training. (Photo: Riccardo Niccoli).*

▲ *F-16B #996 after take-off with the new tail design. (Photo: Major E.G. collection).*

▲ *F-16B #996 Hatzor AB 10.10.2005. (Photo: Sariel Stiller).*

144 Squadron Closing Ceremony - October 10, 2005.

▼ *Four photos by: Sariel Stiller.*

The Defenders of the South Squadron

The 116 Flying Wing Squadron was established on February 7, 1956 as a training squadron for reserve pilots with P-51 Mustangs transferred to Tel Nof AB from 101 Squadron. The squadron became famous at the time thanks to the operation to tear up the telephone lines in the Sinai desert at the start of Operation Kadesh.

On January 15, 1961, the squadron was deactivated. It was reactivated on October 15, 1961. For the second time the squadron was tasked with picking up the 101 Squadron's old aircraft, this time the Dassault Mystère IV. It operated the Mystère IV for about 10 years, participating in the Six Day War and the War of Attrition and made about 1,490 operational sorties, and accomplished in her honor the downing of a Jordanian Hunter. In January 1971, the Mystère IV era in the Flying Wing Squadron ended. Meanwhile, a new era began with the arrival of the first two A-4E Skyhawk jets on February 18, 1971.

The squadron operated from Tel Nof AB until October 3, 1983, the day the squadron was transferred with its aircraft to the new Nevatim AB. In December 2002, the squadron was disbanded, and its aircraft were transferred to the 102 Squadron at Hatzerim AB.

March 31, 2003 - Reactivation ceremony of the squadron with the transfer of 253 Squadron F-16s from Ramon AB to Nevatim AB. The squadron name was changed from the "Flying Wing" to the "Defenders of the South".

The squadron badge was redesigned, and a shadow of an F-16 was added to it. *Lt. Col. Erez Tzelnik*, who deactivated the 253 Squadron, was appointed as the first squadron commander. Squadron 116, in addition to its function as an operational squadron, also functioned as an advanced operational training squadron.

May 28, 2003 - F-16s on an interception standby were launched for the first time following a malfunction that occurred to a Cobra helicopter which was training over Palestinian Authority territory. The helicopter crew managed to land it in the Ella valley and the F-16s safely returned to base.

July 7 - August 14, 2006 - The Second Lebanon War is the Israeli name for the war that took place between IDF forces and Hezbollah in the summer of 2006 in Lebanon and northern Israel.

The war lasted 34 days, although IDF forces continued to operate in Lebanon until October 1. It began with a massive Israeli attack following the kidnapping of IDF soldiers by Hezbollah. During the fighting in Lebanon, the IDF also continued Operation Summer Rains in the Gaza Strip. The squadron participated in the Second Lebanon War and carried out dozens of operational sorties, with nearly half of those operational assault sorties on many targets in Lebanese territory.

December 27, 2008 - January 18, 2009 - Operation Cast Lead. On December 27, in the late morning, the operation began with a wave of heavy air bombardment by the Israeli Air Force using fighter jets and attack helicopters. In the first two bombing waves, 100 targets were hit. On the same day, the IAF attacked another 70 targets in the Gaza Strip. Throughout the operation, a total of 5,400 bombs were dropped by the Air Force. 81% of them were smart bombs, a record in the Air Force's history of air-strikes. 116 Squadron made a significant contribution in this operation and carried out dozens of bombing sorties.

November 11-21, 2012 - Operation Pillar of Cloud or Pillar of Defense. In the course of the operation, the IDF struck more than 1,500 sites in the Gaza Strip, including rocket launchpads, weapon depots and government facilities. Also in this operation, Squadron 116 took a significant part by carrying out dozens of bombing sorties.

August 6, 2013 – Deactivation of 140 Squadron and transfer of its aircraft to 116 Squadron, which thus becomes the largest squadron in the Air Force. The squadron was divided into three units: the operational unit, the operational training course unit, and the advanced operational training course unit. All this with one technical department. One large technical unit was created whose main task was to provide aircraft serviceability for the three units, that flew many missions every day, twice as many as any other squadron in the Air Force. The new and large technical department soon became a "well-oiled" and coordinated outfit that performed all its tasks at the highest level.

July 8 - August 26, 2014 - Operation Protective Edge or Operation Strong Cliff. As the Israeli operation began, the IDF bombarded targets in the Gaza Strip with artillery and air-strikes, Hamas continued to fire rockets and mortar shells into Israel in response. The Air Force attacked thousands of targets inside Gaza strip and 116 Squadron contributed significantly n this operation, alongside its activity as a training squadron for fresh F-16 pilcts.

May 2015 - The squadron operational activity has come to an end, but the Advanced Operational Training Course Division continued to operate the F-16 until December 31, when the squadron was deactivated. The last squadron commander was Lt.Col. Yotam.

▲ *Negev Squadron F-16s Reception Ceremony by the new 116 Squadron. (Photo: Major E.G. collection).*

▼ *The commander of 253/116 Squadron replaces the emblem of the 253 Squadron with that of 116 Squadron. (Photo: Major E.G. collection).*

PHOTO GALLERY

Photo: Alberto Mocchetti

▲ F-16A #100 and #102 taxiing for take-off. Nevatim AB - May 15, 2015. ▼

Photo: Riccardo Niccoli

Photo: Riccardo Niccoli

A series of photographs of F-16A #107 the world's leading F-16 MiG Killer with 6.5 victories.

(Photos: Dubi Tal - Albatross Aerial Photography)

▼ During take-off. Nevatim AB - November 29, 2013. (Photos: Moni Shafir).

▲ *Just landed - Nevatim AB April 21, 2013. (Photo: Hezi Shmueli).*

▲ *Close-up - December 28, 2010. (Photos: Moni Shafir).*

▲ *Perspective - April 15, 2013. (Photos: Moni Shafir).*

A series of beautiful aerial photographs taken on June 22, 2006.
(Photos: Major Ofer)

149

▲ F-16A #111 during take-off - Ovda AB April 21, 2013. (Photo: Hezi Shmueli).

▲ F-16A #111 - during operation "Cast Lead" loaded with two M117 General-Purpose (GP) bombs.

▲ F-16A #111 - during operation "Cast Lead" prepared for a night sortie.

▲ F-16A #243 taxiing for take-off. (Photo: Riccardo Niccoli).

▲ F-16A #112 - IAF pilot gives the OK on his way for take-off. (Photo: Major E.G. collection).

▲ F-16A #112 parking inside its protective shelter. (Photo: Riccardo Niccoli).

▲ F-16A #112 during take-off - Ovda AB April 21, 2013. (Photo: Hezi Shmueli).

▲ "Netz" #116 pilot cleared to taxi by the ground crew chief, for an operational mission, armed with four live AIM-9L missiles. Nevatim AB. (Photo: IDF/AF).

▲ F-16A #116 on its way for a training sortie loaded with Mk-82 inert bombs. (Photo: Major E.G. collection).

▲ F-16A #116 parked inside its protective shelter. (Photo: Sariel Stiller).

▲ F-16A #116 transported to its last destination, after it was cannibalized for parts. (Photo: Major E.G. collection).

▲ F-16A #126 - Photo from October 25, 2004 with the old 116 Squadron badge carrying Pave Penny laser designating pod "Mehuspas".

▲ F-16A #126 used as "Red" or Aggressor. (Photo: Amit Agronov).

▲ F-16A #126 during a formation flight over the Negev.

▲ F-16A #129 during take-off for a training mission. (Photo: Lior Kestner).

▲ F-16A #129 last checks before take-off. (Photo: Major E.G. collection).

▲ F-16A #129 landing back home safely. Note the Elta EL/L-8212 ECM pod at the centreline. (Photo: Sariel Stiller).

▲ F-16A #131 in first daylight, parked ready and waiting with information for its pilots. (Photo: Major E.G. collection).

▲ F-16A #131 lifting from the runway at takeoff. (Photo: Lior Kestner).

F-16A #131 - A series of photos show the process of ground crew preparing a plane to flight. (Photos: Sariel Stiller).

▲ F-16A #135 weight on wheels upon landing. (Photo: Lior Kestner).

F-16A #138 - transported to its last destination, after it was cannibalized for parts. (Photo: Major E.G. collection). ▶

▲ F-16A #232 during maintenance.
(Photo: Major E.G. collection).

▲ F-16A #242 during maintenance (Photo: Major E.G. collection).

▲ F-16A #243 air refueled by a IAI modified Boeing 707 tanker. (Photo: Moni Shafir).

▲ F-16A #243 taxiing for take-off. (Photo: Alberto Mocchetti).

▲ F-16A #246 with an AN/ALQ-119 ECM pod but without the squadron badge and number on its tail.

▲ ▼ Photo from January 3, 2009. F-16A #249 during operation "Cast Lead" loaded with four M117 General-Purpose (GP) bombs. (Photos: Major E.G. collection).

▲ F-16A #255 during a static exhibition. (Photo: Moni Shafir).

▲ F-16A #255 taxiing for take-off. (Photo: Alberto Mocchetti).

▲ F-16A #255 with F-16B #996 awaiting take-off permission. (Photo: Riccardo Niccoli).

▲ F-16A #260 - Photo from February 10, 2010 during deployment at Ovda AB. (Photo: Major E.G. collection).

▲ F-16A #261 taxiing to the squadron after landing. (Photo: Hezi Shmueli).

▲ F-16A #265 after take-off during deployment at Ovda AB. (Photo: Hezi Shmueli).

▲ F-16A #272 during a static display. (Photo: Udi Burg).

▲ F-16A #273 in display during a static exhibition.

▲ F-16A #275 and #138 escorting the first two F-16I "Sufa" fighter jets that arrived in Israel on February 19, 2004.
(Photo: Major E.G. collection).

▲ Photo from October, 2012 formation of four F-16s during a take-off. (Photo: Lior Kestner).

▲ Photo from June, 2012 some of the squadron planes line. (Photo: Major E.G. collection).

▲ June 8, 2010 ground crew loading a Netz with M117 bombs for an operational sortie. (Photo: Major E.G. collection).

▲ Photo from January 3, 2009. F-16A #282 during operation "Cast Lead". (Photo: Major E.G. collection).

▲ F-16A #282 taxiing to the squadron after landing. (Photo: Hezi Shmueli).

▲ Photo from January 3, 2009. F-16A #284 during operation "Cast Lead". (Photo: Major E.G. collection).

▲ F-16A #284 return from a training flight. (Photo: Moni Shafir).

▲ F-16A #285 inside its protective shelter, photo from June 2012. (Photo: Major E.G. collection).

▲ F-16A #292 and #112 during a "hot refueling" exercise. (Photo: Major E.G. collection).

▲ Ground crew preparing F-16A #292 for an operational sortie. (Photo: Major E.G. collection).

▲ F-16A #292 at last point, waiting permission for take-off. Note the EL/L-8212 ECM pod at the centreline station, and a Mk.82 training water bomb under the wing. (Photo: Major E.G. collection).

▲ F-16A #298 at Ovda AB scrapyard. (Photo: Major E.G. collection).

▲ *F-16A #728 now in the IAI scrapyard. (Photo: Leah Khananashvil)*

▲ *F-16A #729 - (Photo: Marco Pennings).*

▲ *F-16A #737 now in the IAI scrapyard. (Photo: David Weinrich)*

▲ F-16A #754 used for firefighting and rescue practice at Ovda AB.

▲ F-16A #766 in static display at IAF Museum - Hatzerim. (Photo: Tzahi Ben-Ami).

▲ F-16A #769 and #138 positioned as dummy planes at Nevatim AB. (Photo: Major E.G. collection).

▲ F-16B #004 during deployment to Turkey - November 2003.

▲ F-16B #004 Photo from July 2007 - Nevatim AB. (Photo: Major E.G. collection).

▲ F-16B #004 after take-off for a training mission. (Photo: Moni Shafir)

▲ F-16B #004 after take-off for a training mission. (Photo: Lior Kestner)

▲ F-16B #004 just landed. (Photo: Sariel Stiller).

▲ F-16B #004 taxiing to the squadron back from a training mission. (Photo: Sariel Stiller)

▲ *F-16B #017 on display during a static exhibition.*

▲ *F-16B #017*

▲ *F-16B #017 during deployment to Turkey - November 2003.*

▲ F-16B #974 taken while landing, (Photo: Tzahi Ben-Ami).

▲ F-16B #974 flying over Turkish countryside during deployed to Turkey - November 2005.

▲ F-16B #979 taxiing for take-off. (Photo: Sariel Stiller)

▲ *F-16B #979 upon landing, (Photos: Sariel Stiller)* ▼

▼ *F-16B #979 receives a permit to roll from a ground crew member. (Photo: Sariel Stiller)*

▲ *F-16B #981 taxiing for take-off. (Photo: Riccardo Niccoli).*

▲ *F-16B #979 at the closing ceremony of Squadron 144 and the transfer of its aircraft to the Nevatim AB.* ▼
(Photos: Sariel Stiller)

▲ F-16B #983 now in the IAI scrapyard. (Photo: Leah Khananashvil).

▲ F-16B #989 taxiing for take-off August 5th, 2013. (Photo: Moni Shafir).

▲ F-16B #989 taxiing for take-off May 17, 2015. (Photo: Riccardo Niccoli).

▲ *F-16B #993 taken while landing on November 29, 2013. (Photo: Moni Shafir).*

▲ *F-16B #993 taxiing to the squadron after landing. (Photo: Hezi Shmueli).*

▲ *F-16B #979 taxiing for take-off. (Photo: Sariel Stiller)*

▲ *F-16B #995 taxiing for take-off May 17, 2015. (Photo: Riccardo Niccoli).*

▲ *F-16B #996 in the process of landing. October 29, 2012. (Photo: Lior Kestner).*

▲ *F-16B #996 taxiing to its Hardened Shelter. October 29, 2012. (Photo: Lior Kestner).*

November 15-25, 2010 - Italian and Israeli jets deployed to Decimomannu air base in the island of Sardinia for the bi-national exercise VEGA 2010. Two F-16Bs from the Defenders of the South Squadron participate in the exercise (#004 and #993).

Photo: Massimo Pieranunzi

Photo: Major E.G. collection

Photo: Major E.G. collection

Photo: Giovani Colla

Photo: Major E.G. collection

Photo: Giovani Colla

Photo: Giovani Colla

Photo: Giovani Colla

Photos: Major E.G. collection

Photo: Major E.G. collection

Photo: Massimo Pieranunzi

Photo: Massimo Pieranunzi

Photos: Major E.G. collection

Photos: Major E.G. collection

187

October 26, 2012 - Last flight and farewell ceremony of Brigadier General (ret.) Rafi Berkovich (Berko). Berko was the first F-16 pilot to shoot down an enemy aircraft (Mil-8) and the only pilot ever to shoot down 3 jets in 45 seconds. He downed two MiG-23s with AIM-9Ls and a MiG-21 with the F-16's cannon. He chose to fly his last flight on the same F-16... #116.

116 Squadron Closing Ceremony December 7th, 2015 (Photos: Major E.G. collection).

Photo: Major Ofer

The Flying Dragon Squadron

The 115 Flying Dragon was established in July 1954 as a Mosquito PR-16 reconnaissance unit within the 109 Squadron at Hatzor AB. In May 1956, the unit was declared a squadron, and in July it was transferred to Ekron (now Tel-Nof) AB. After Operation Kadesh the squadron was deactivated. It was reactivated on March 20, 1969 as the third A-4 Skyhawk squadron in the IAF. In February 1985, the squadron moved its aircraft to the new Nevatim AB. The squadron disbanded on July 21, 1994, with some of its aircraft going into storage while others were dispersed among remaining IAF A-4 Squadrons. In March 2005, the squadron was reformed at Ovda AB as an advanced training unit for aerial warfare. Lieutenant Colonel *Amnon El-Dar* was appointed squadron commander. The F-16s operate as a "red force" (aggressors) to train fighter pilots in combat theory in line with enemy doctrines. The "Flying Dragon", known also as the "Red Squadron", is not an ordinary squadron.

Acting as the hostile force while training with the various IAF's units, the 115 "Flying Dragon" squadron's main objective is to simulate how the enemy thinks and operates, increasing the IAF's capabilities and wartime readiness significantly. The aircrew members of the aggressor squadron even use foreign name tags and distinctive red ranks to mimic the enemy. It is not an operational unit, though all its pilots have emergency postings and its F-16s are equipped to serve as combat aircraft in the event of war under 116 Squadron command.

Modeled on USAF aggressor squadrons, the unit is also offering its services to other nations. In May 2006 it trained with the Massachusetts Air National Guards 101st Fighter Squadron and in 2008 the squadron provided desert training for 55 Czech Air Force pilots prior to their deployment to Afghanistan. As a tribute to Czechoslovak military assistance to Israel during the 1948 Arab-Israeli War, the training session was named "Etzion", once the codename for the Czech airfield at Žatec from where a great deal of aircraft and material were dispatched to Israel. In early December 2010, 115 Squadron hosted Italian Air Force Panavia Tornado's at Ovda AB, conducting a week-long joint training session. In December 2011, the Israeli and Italian Air Forces completed a two-week joint training exercise. The exercise involved pilots flying F-16As, F-16Cs and F-15Is from three Israeli squadrons, pitted against Italian Air Force pilots flying Eurofighter Typhoons and Panavia Tornado strike fighters. In March 2012, the Polish Air Force's 10th Tactical Squadron deployed to Ovda for a two-week joint exercise with Israel's 115, 117 and 106 squadrons. In November 2013, the Blue Flag exercise was held. The exercise spanning over two weeks began with a week-long orientation for the visiting units; eight US F-15Es from the 492nd Tactical Fighter Squadron from Lakenheath/UK, four Hellenic Air Force F-16Es from the 340 Squadron ("Asteri") and Italian Tornado and AMX strike fighters.

On December 31, 2016, 115 Squadron retired its last F-16As, replacing them with F-16Cs. This was the final retirement of the Netz in the Israel Air Force.

Photo: Major Ofer

PHOTO GALLERY

▲ F-16A #113. (Photo: Hezi Shmueli). ▼ ▼

▲ F-16A #124 taxiing for take-off. (Photo: Lior Kestner).

▲ F-16A #124 touch-down. (Photo: Major E.G. collection).

▲ F-16A #124 taxiing back to the squadron. (Photo: Hezi Shmueli).

▲ ▼ ▼ *F-16A #124 and #223 exhibited in the squadron during an open day. (Photo: Major E.G. collection).*

▲ F-16a #216 taken on May 17, 2015 while taxiing back from a training mission. (Photo: Riccardo Niccoli).

▲ *F-16A #220 taxiing out from its shelter for take-off. (Photo: Major E.G. collection).*

▲ *F-16A #220 during take-off. (Photo: Hezi Shmueli).*

▲ *F-16A #220 carries an AN/ALQ-119 ECM pod climbing after take-off (Photo: Hezi Shmueli).*

▲ *F-16A #223 with F-15C from 106 squadron on their way for a training sortie. (Photo: Major E.G. collection).*

▲ *F-16A #223 is waiting for the green light for take-off on a training sortie. (Photo: Major E.G. collection).*

▲ *F-16A #223 with its formation on their way for a training sortie. (Photo: Major E.G. collection).*

▲ F-16A #233 at touchdown. (Photo: Major E.G. collection).

▲ F-16A #234 taxiing out its Hardened Shelter at Ovda AB. (Photo: Major E.G. collection).

▲ F-16A #234 taxiing to the take-off point. (Photo: Alberto Mocchetti).

▲ *Photo from January 3, 2009. F-16A #234 during operation "Cast Lead". (Photo: Major E.G. collection).*

▲ *F-16A #234 during take-off. (Photo: Hezi Shmueli).*

▲ *F-16A #234 with its 3 kill marks after take-off. (Photo: Major E.G. collection).*

▲ *F-16A #239 on its way as "aggressor" waiting its turn to take-off. (Photo: Major E.G. collection).*

▲ *F-16A #239 at "last point" checked up by ground crew member. (Photo: Major E.G. collection).*

▲ *F-16As' parking in the squadron line. (Photo: Major E.G. collection)*

▲ *F-16A #250 carrying two kill marks is directed to its parking point. (Photo: Major E.G. collection).*

▲ *F-16A #250 with its formation on their way for a training sortie. (Photo: Major E.G. collection).*

▲ *F-16A #260 on a static display at Hatzerim AB.*

201

▲ *F-16A #260 climbing after take-off, note the 60 years logo applied under the fin. (Photo: Hezi Shmueli).*

▲ *F-16A #264 on its way to routine training mission. (Photo: Major E.G. collection).*

▲ *F-16A #273 seen here at Hatzor AB. (Photo: Sariel Stiller).*

▲ ▼ *F-16a #115 (ex #275) taken while taxiing back from a training mission. (Photo: Riccardo Niccoli).*

▲ *F-16A #275 taxiing for take-off from Hatzor AB. (Its # was change to 115 on 18.12.2012) (Photo: Sariel Stiller).*

▲ *F-16A #275 taxiing for take-off from Hatzor AB. (Photo: Sariel Stiller).*

▲ *F-16A #281 awaiting tower permission for take-off. (Photo: Hezi Shmueli).*

▲ *F-16A #281 just after take-off climbing out. (Photo: Hezi Shmueli).*

▲ F-16A #281 carries an AN/ALQ-119 ECM pod taxiing back to its Hardened Shelter at Ovda.
(Photo: Major E.G. collection).

▲ F-16A #281 after take-off, red tank and missiles were used to mark the aggressor aircraft in combat training.
(Photo: Hezi Shmueli).

▲ Four F-16As waiting for a new training mission. (Photo: Major E.G. collection).

▲ F-16A #285 pilot cleared to taxi by the ground crew chief. (Photo: Hezi Shmueli).

▲ F-16A #285 during take-off. (Photo: Hezi Shmueli).

▲ F-16A #296 and #220 taxiing for take-off. (Photo: Major E.G. collection).

▲ F-16A #296 pilot waving goodbye to the photographer. (Photo: Major E.G. collection).

▲ F-16A #296 climbing on full military power after take-off. (Photo: Hezi Shmueli).

▲ F-16A #296 taken after landing. (Photo: Major E.G. collection).

▲ F-16A #296 ready for long-term storage after being taken out of service. (Photo: Major E.G. collection).

A series of beautiful aerial photographs taken on November, 2013, during the Blue Flag exercise. The exercise spanning over two weeks began with a week-long orientation for the visiting units: eight US F-15Es from the 492ns Tactical Fighter Squadron from Lakenheath, UK, four Hellenic Air Force F-16Es from the Hellenic Air Force 340 Squadron ("Asteri") and Italian Tornado and AMX strike fighters (Photos: Major Ofer)

211

212

IAF Flight Test Center Squadron
MANAT

Upon arrival of the F-16 jets in the IAF, an application was submitted to the Air Force Headquarters by TFC Commander Colonel *Yossi Ron* and Chief Engineer *Yair David* to dedicate one aircraft for test activities. The Air Force headquarters objected to the issue, but after the intervention and recommendation of Brigadier General *Avihu Ben-Nun*, at the time the commander of Tel Nof AB, the Air Force commander, Major General *David Ivri* approved a limited budget to the cause. *Yair David* traveled to the General Dynamics plant to prepare a joint characterization proposal for approval by the US Air Force Headquarters. The Americans offered a separate aircraft for each type of experiment. But within the budget limits approved, the IAF engineers had to make a list of parameters that would be tested in the future experiments and combine all the sensors in one aircraft.

F-16A C/N 6V-67 from block 10D was provided with all the necessary sensors and systems and the aircraft received number 299 in the Israeli Air Force. The jet arrived in Israel in October 1982 and the inauguration flight was performed by the late test pilot *Moti Rader*.

In the mid-1990's, after the removal of the IAI Kfir aircraft from service, which were used amongst other things as chase aircraft, F-16B #993 was transferred from 144 Squadron to the TFC/MANAT squadron for use as a chase aircraft. The aircraft was returned to its original squadron in the end of 2000.

F-16A #299 was in active service with the TFC till 2012.

▼ *Flight Test Center Squadron F-16A #299 during first test flight carrying four MK-84 bombs (1985).*

▲ *F-16A #299 shown here with the first FTC squadron insignia and an orange chevron on the leading edge of its tail on static display in 1988 during an open day at Tel Nof AB. note the IMI VER-4 bomb rack for four bombs, tested but never introduced into service.*

F-16A #299 shown here on static display during IAF 50th anniversary exhibition. ▶

▲ *Flight Test Center Squadron F-16B #993 in a formation flight over Israel. (Photos: IAF Magazine)* ▲

◀ *Flight Test Center Squadron F-16B #993 after it was painted in the standard IAF F-16s camouflage. (Photo: Major E.G. collection).*

▼ *Flight Test Center Squadron F-16B #993 photo taken at Tel Nof in May 2000. (Photo: Riccardo Niccoli).*

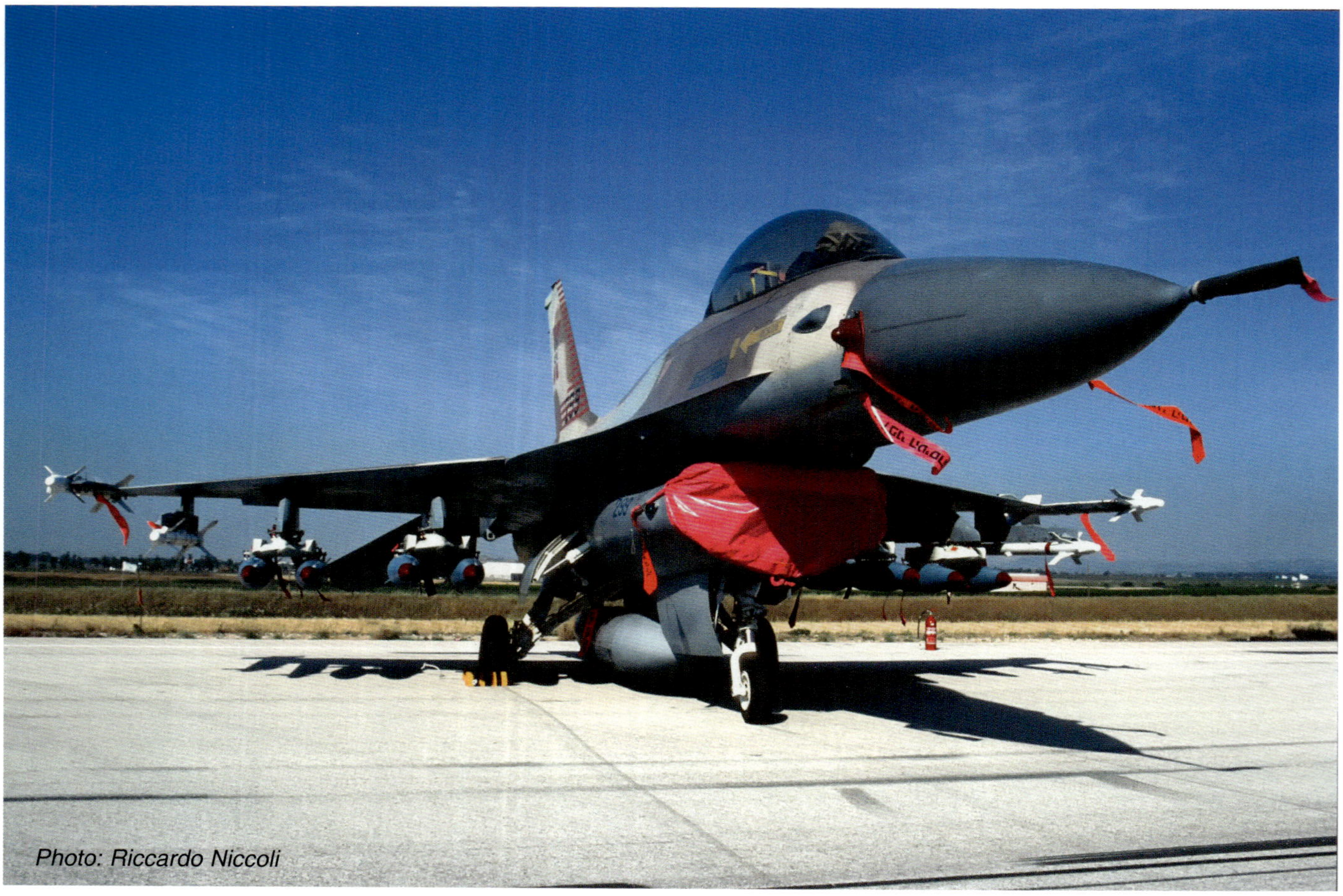
Photo: Riccardo Niccoli

▲ *F-16A #299 shown here on static display in its tail markings applied since 1990 - Tel Nof AB 2000.* ▼

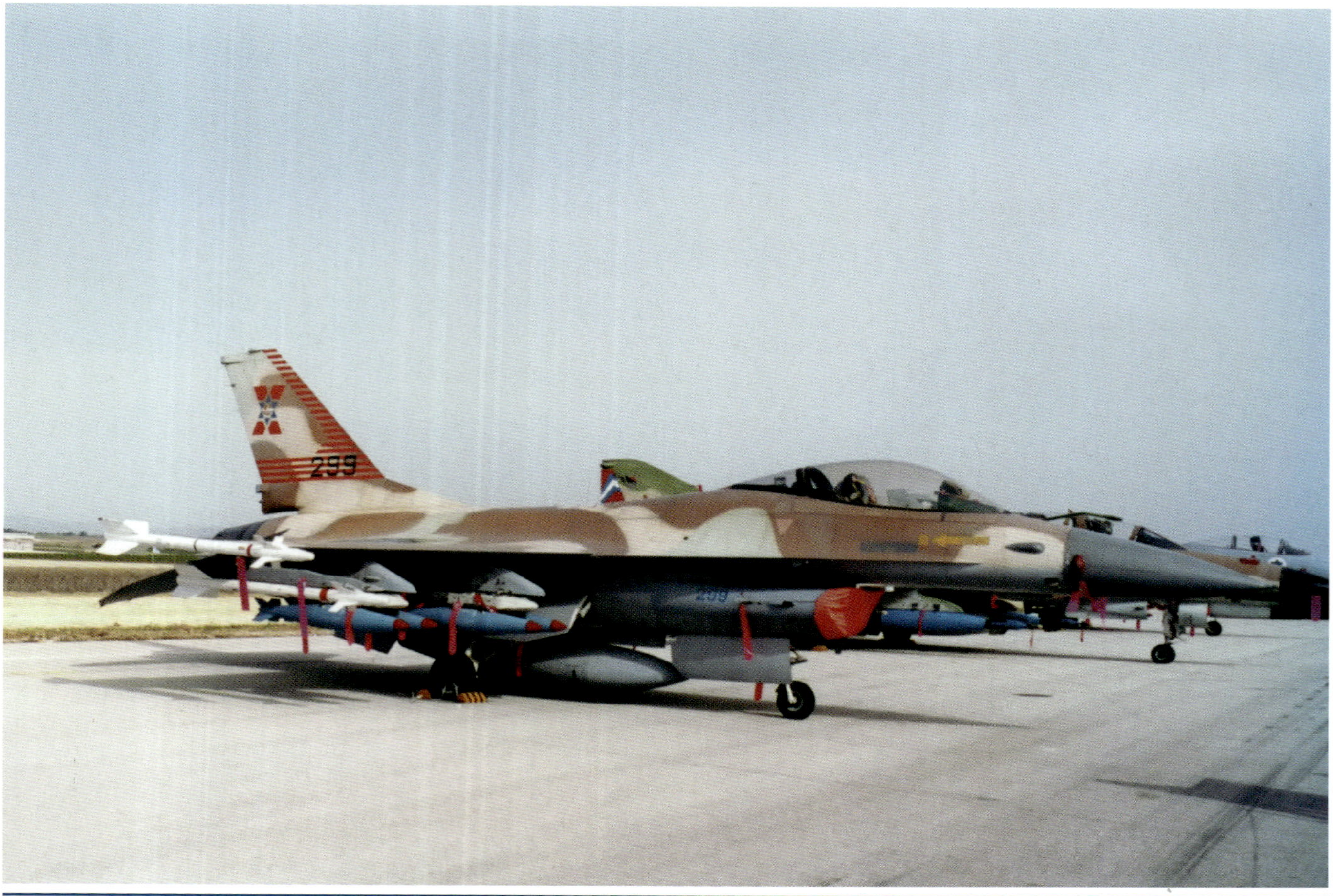
299

F-16 A/B Technical Specifications

Length:	49.6 ft (15.1m)
Height:	16.52 ft (5.01m)
Wing Span:	31 ft (9.45 m)
Wing area:	300 sq. Ft. (27.87 mq.)

Engine:
Pratt & Whitney F100-PW-100 turbofan engine delivering 25,000 lb (111.2 kN) of thrust with afterburner;

Weight:
Empty F-16A:	14,567 lb (6,607 kg)
Empty F-16B:	15,141 lb (6,868 kg)
Maximum at take-off:	35,400 lb (16,057 kg)

Performance:
Max level speed (at 40,000 ft): Mach 2.0+ (2,124 km/h or 1,320 mph or 1,146 kts).
Ceiling: 55,000 ft (18,290 m)
Radius of action: 500+ nm (925 km, or 575 miles)
Ferry Range with external tanks : 2,100 nm (3,890 km, or 2,415 miles)

Fuel system:
Internal F-16A:	6,972 lb (3,162 kg)
Internal F-16B:	5,785 lb (2,624 kg)

External fuel tanks: up to two 370 US Gal (1,400 lt) under wing tanks, plus one 300 US Gal (1,136 lt) at the ventral station. Later, also two 600 US Gal (2,271 lt) under wing tanks designed by IMI.

Armament:
Air-to-air:
M61A1 Vulcan six-barrel 20mm gun with 500 rounds plus up to six AIM-9P3/L, M Sidewinder IR-guided missiles

Air-to-ground:
A large payload of US and Israeli munitions, such as Mk.82, Mk.83, Mk.84 bombs, M117 bombs, Mk.20 Rockeye cluster bombs AGM-65 Maverick missiles. After 1992, also laser-guided bombs such as GBU-10, GBU-12, GBU-16, Griffin, associated to the AN/AAS-35 Pave Penny targeting pod.

Maximum external loads (F-16A): 15,200 lb (6,894 kg).

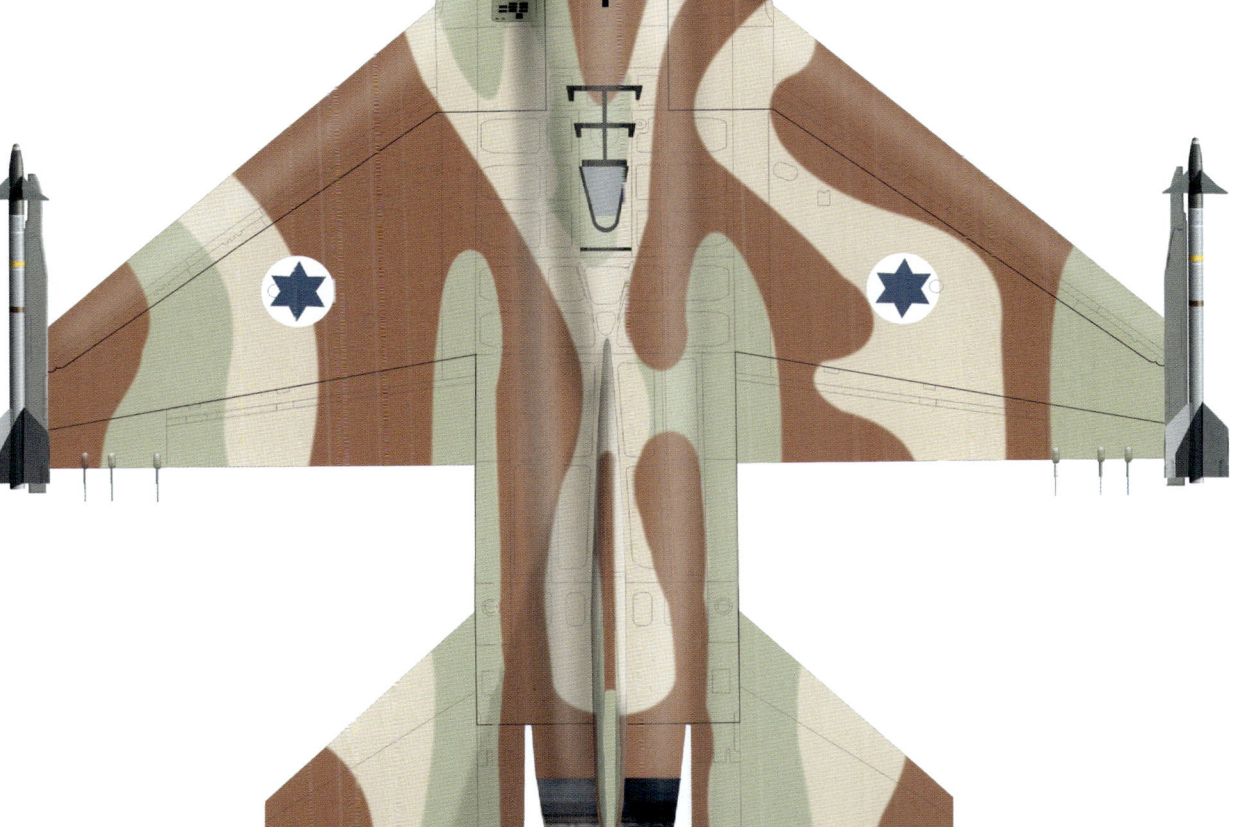

Israel Aircraft Industries and Elbit
ACE PROJECT

In the late 1990s, Israel Aircraft Industries and Elbit thought that it was time to present the IDF/AF with a robust upgrading program for the Netz fleet. The project was called ACE (Avionics Capability Enhancement), and was also internationally introduced at the Paris Le Bourget salons in 1999 and 2001, using as prototype the F-16B serial '986' (ex-USAF 78-0111) which was assigned to the industries in December 1998, receiving the civil registration "4X-ACE". The new equipment to be integrated included the IAI/Elta EL/M-2032 radar, Elbit mission computers, and MFD displays, with a new HUD, DASH helmet, Elta-Elisra electronic self-defence unit, and a data-link. The new weaponry included the Rafael Litening navigation and targeting pod, Derby and Python 4 air-to-air missiles, and the Spice, Ofer and Popeye 2 precision guided bombs. The first flight of the prototype occurred on June 5, 2001. However, the Netz ACE program was too expensive for the IDF/AF, and was discarded, also because the air force was already committed to the F-16I Sufa program, which was requesting huge funds. The Netz fleet was thus maintained in service 'as it was', also in consideration of its main training mission.

Photo: IAI via Riccardo Niccoli

F-16A #107 named "Sufa" (Storm) - before the First Lebanon War (Operation Peace for Galilee).

F-16A #138 named "Shahak" (Sky) credited with one kill mark on it. (Ami Lustig 10.6.1982 Syrian MiG 23)

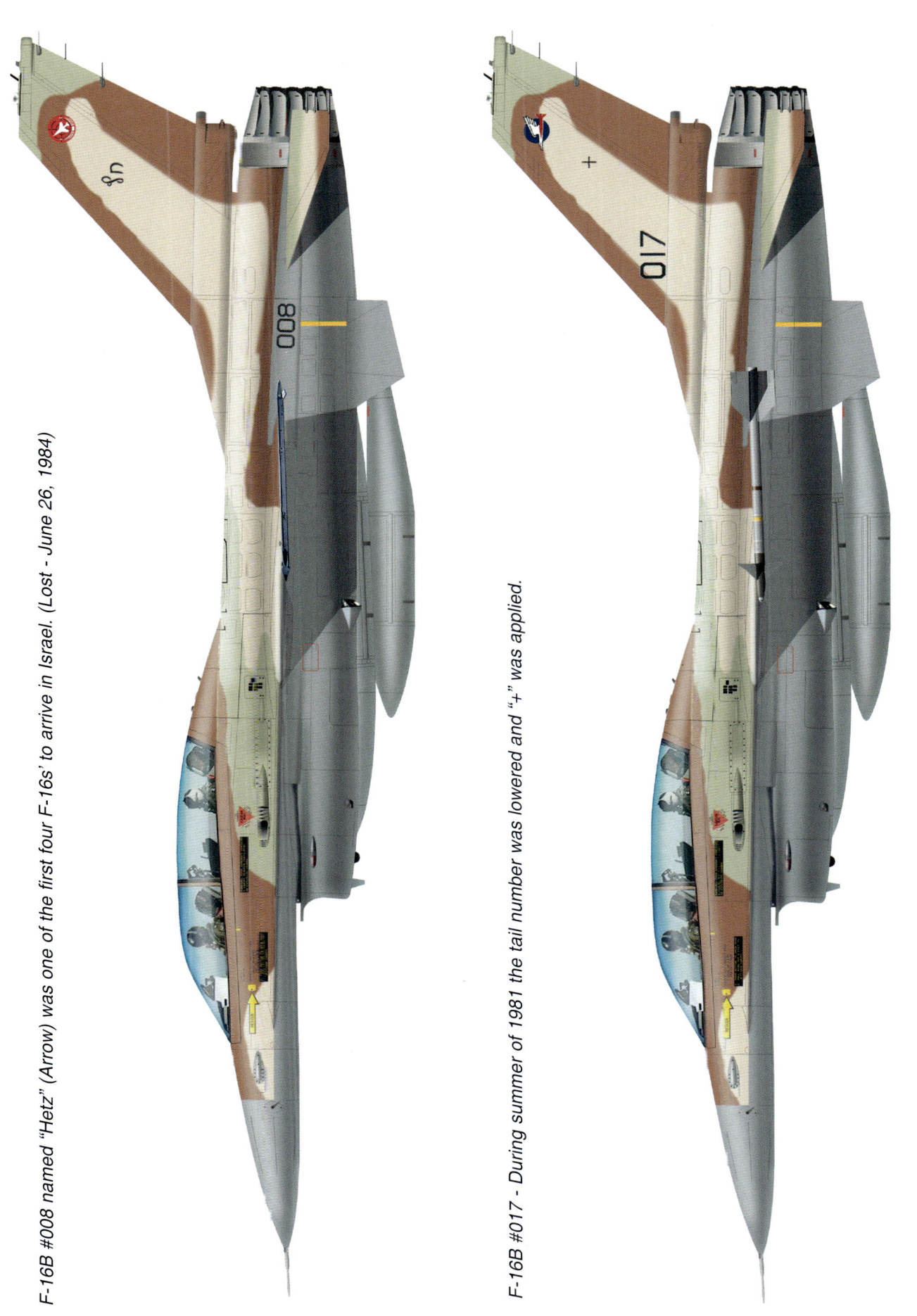

F-16B #008 named "Hetz" (Arrow) was one of the first four F-16s' to arrive in Israel. (Lost - June 26, 1984)

F-16B #017 - During summer of 1981 the tail number was lowered and "+" was applied.

221

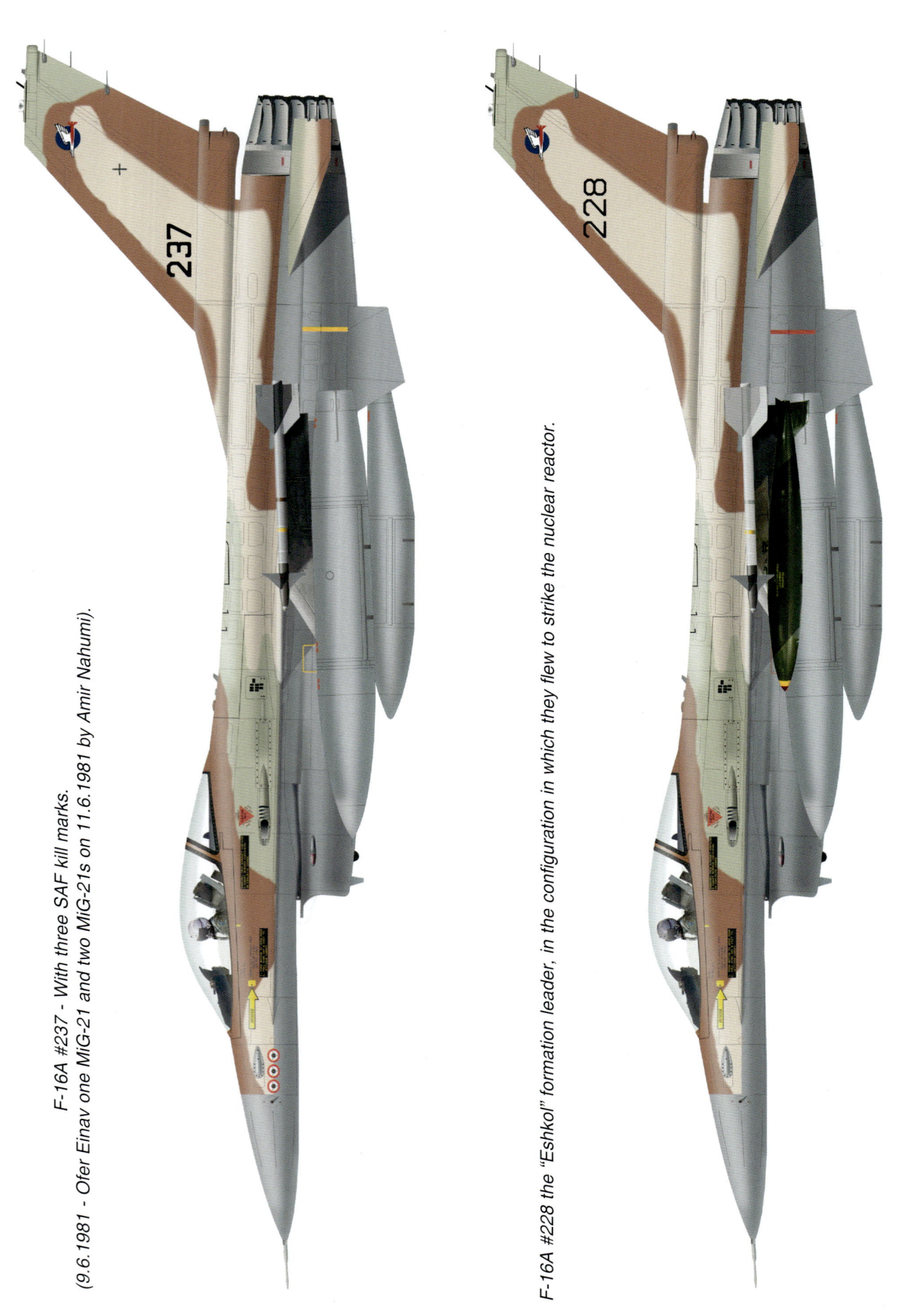

F-16A #237 - With three SAF kill marks.
(9.6.1981 - Ofer Einav one MiG-21 and two MiG-21s on 11.6.1981 by Amir Nahumi).

F-16A #228 the "Eshkol" formation leader, in the configuration in which they flew to strike the nuclear reactor.

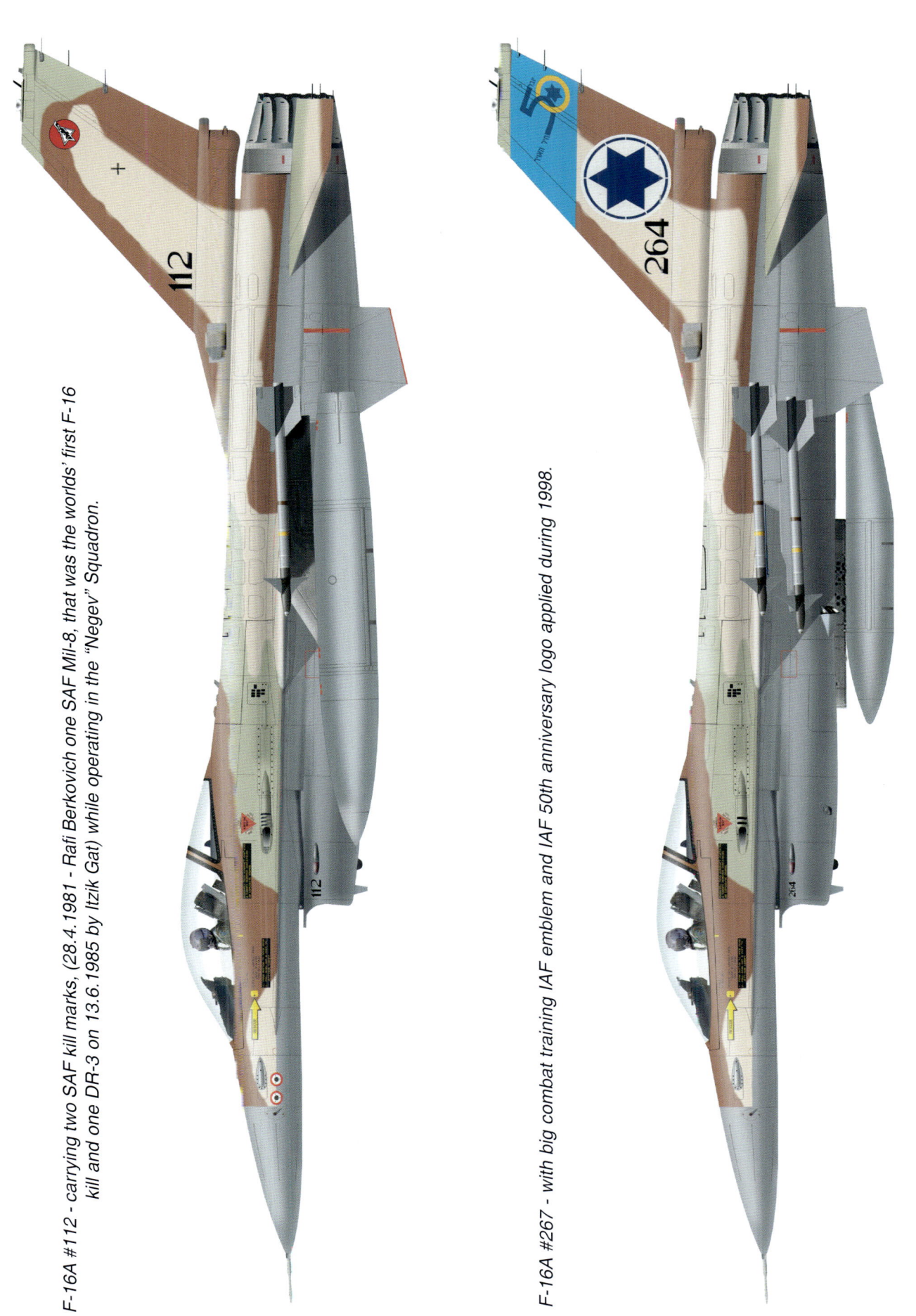

F-16A #112 - carrying two SAF kill marks, (28.4.1981 - Rafi Berkovich one SAF Mil-8, that was the worlds' first F-16 kill and one DR-3 on 13.6.1985 by Itzik Gat) while operating in the "Negev" Squadron.

F-16A #267 - with big combat training IAF emblem and IAF 50th anniversary logo applied during 1998.

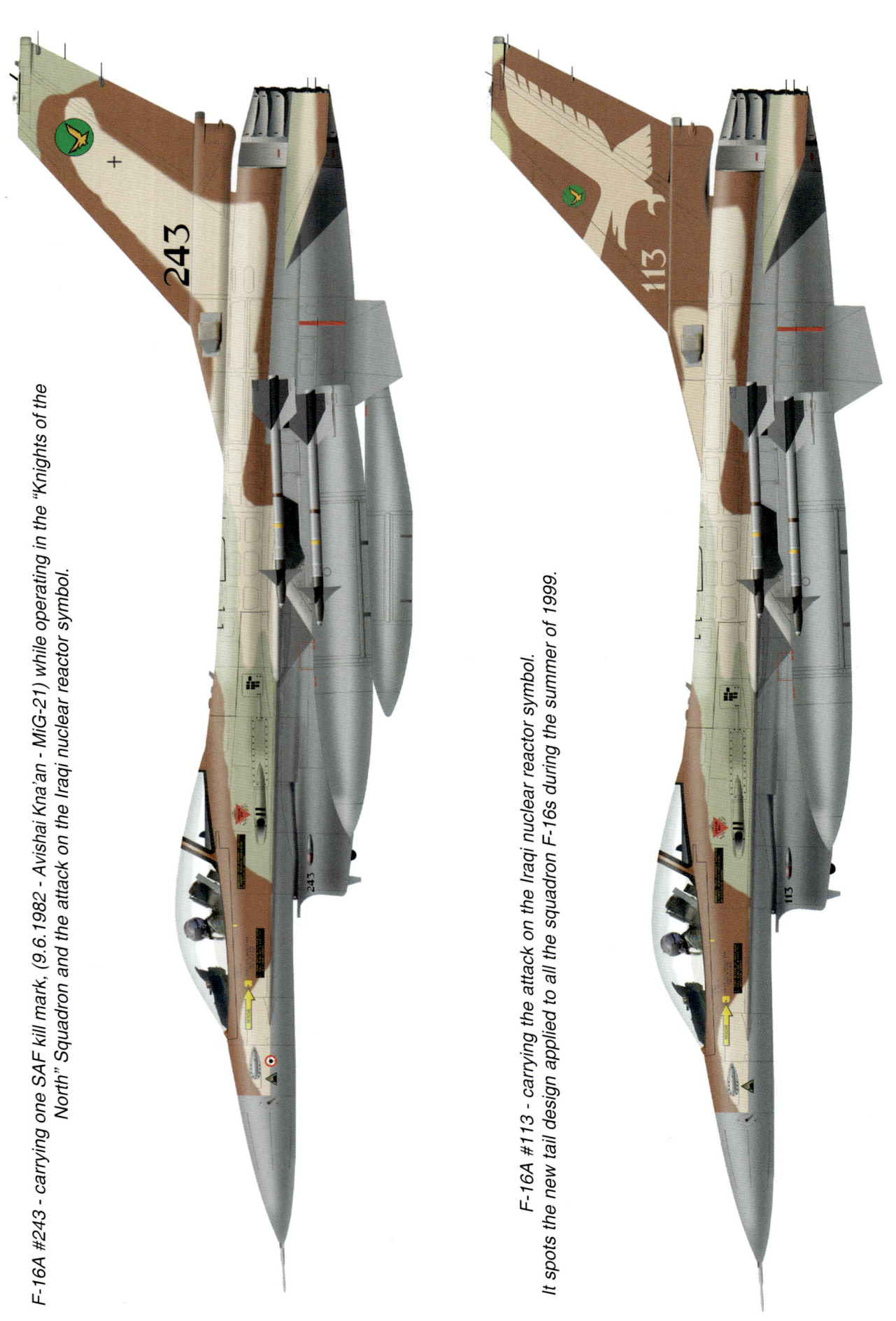

F-16A #243 - carrying one SAF kill mark, (9.6.1982 - Avishai Kna'an - MiG-21) while operating in the "Knights of the North" Squadron and the attack on the Iraqi nuclear reactor symbol.

F-16A #113 - carrying the attack on the Iraqi nuclear reactor symbol. It spots the new tail design applied to all the squadron F-16s during the summer of 1999.

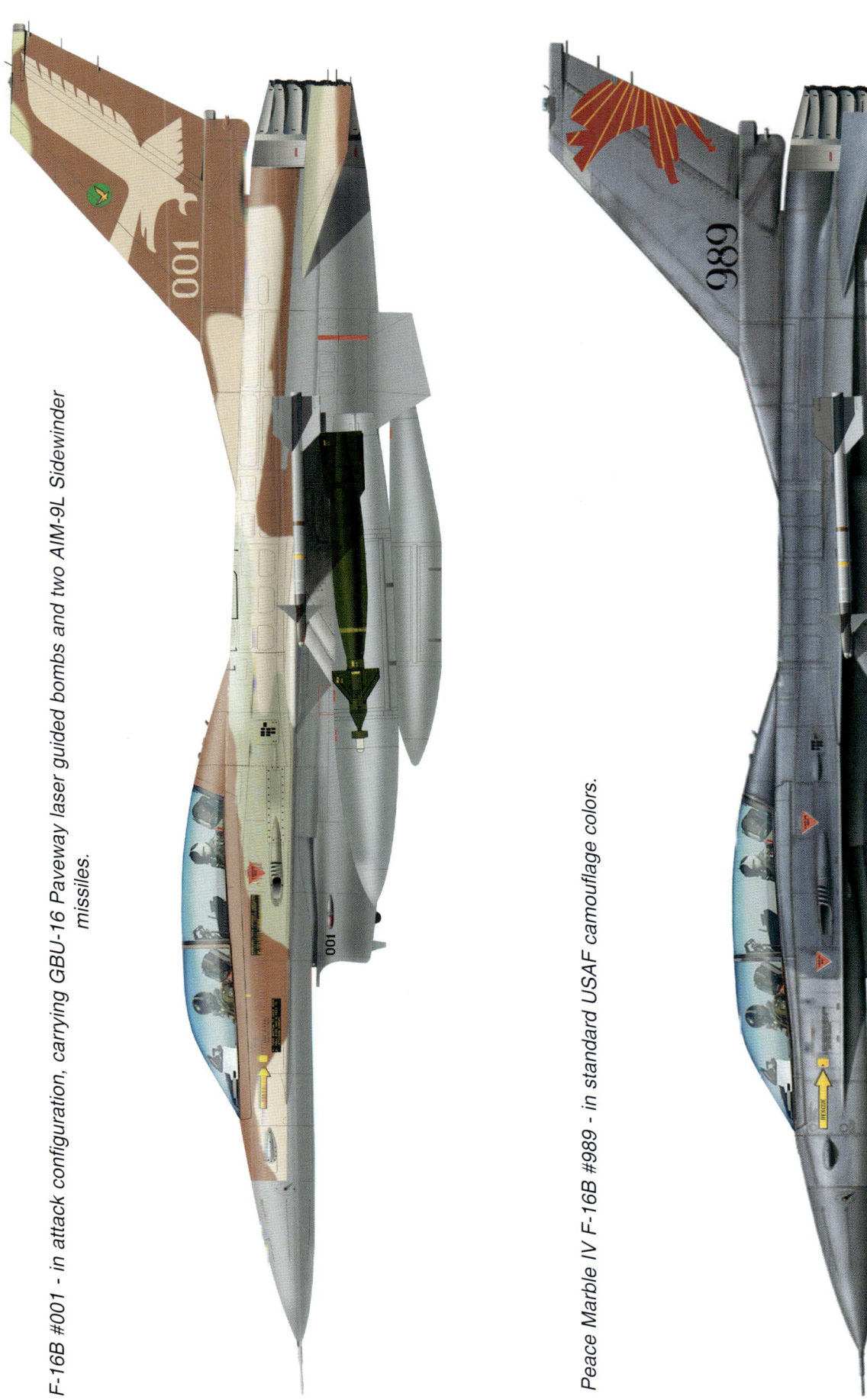

F-16B #001 - in attack configuration, carrying GBU-16 Paveway laser guided bombs and two AIM-9L Sidewinder missiles.

Peace Marble IV F-16B #989 - in standard USAF camouflage colors.

Peace Marble IV F-16A #755 - in standard IAF camouflage colors painted in 1998.

Peace Marble IV F-16A #757 - in standard IAF camouflage colors painted in 1998. With the new tail design art applied to all the squadron F-16s from summer of 2001.

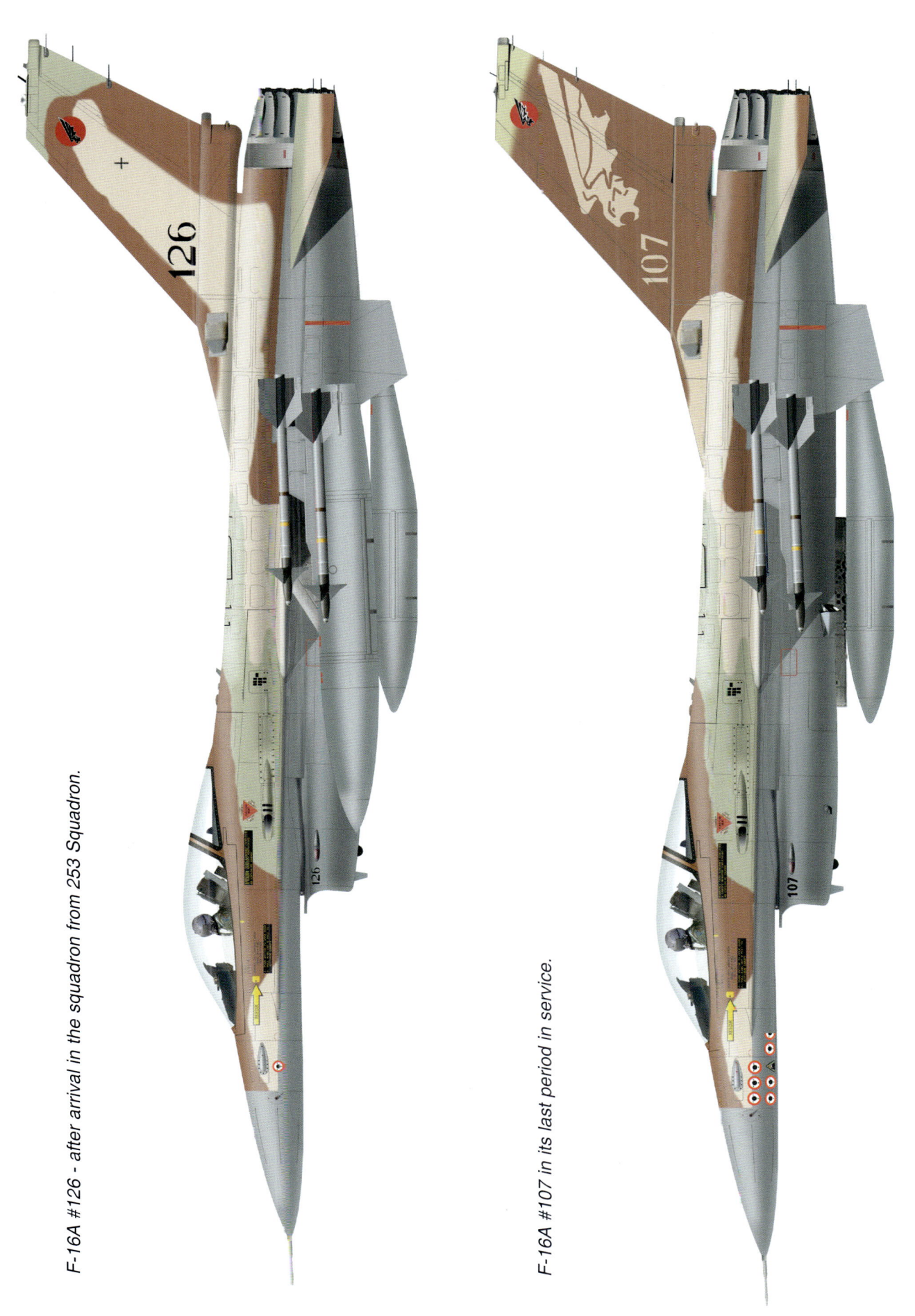

F-16A #126 - after arrival in the squadron from 253 Squadron.

F-16A #107 in its last period in service.

F-16B #993 - in attack configuration, carrying GBU-16 Paveway laser guided bombs and two AIM-9L Sidewinder missiles.

F-16A #273 - with the 60 years anniversary design applied under the fin.

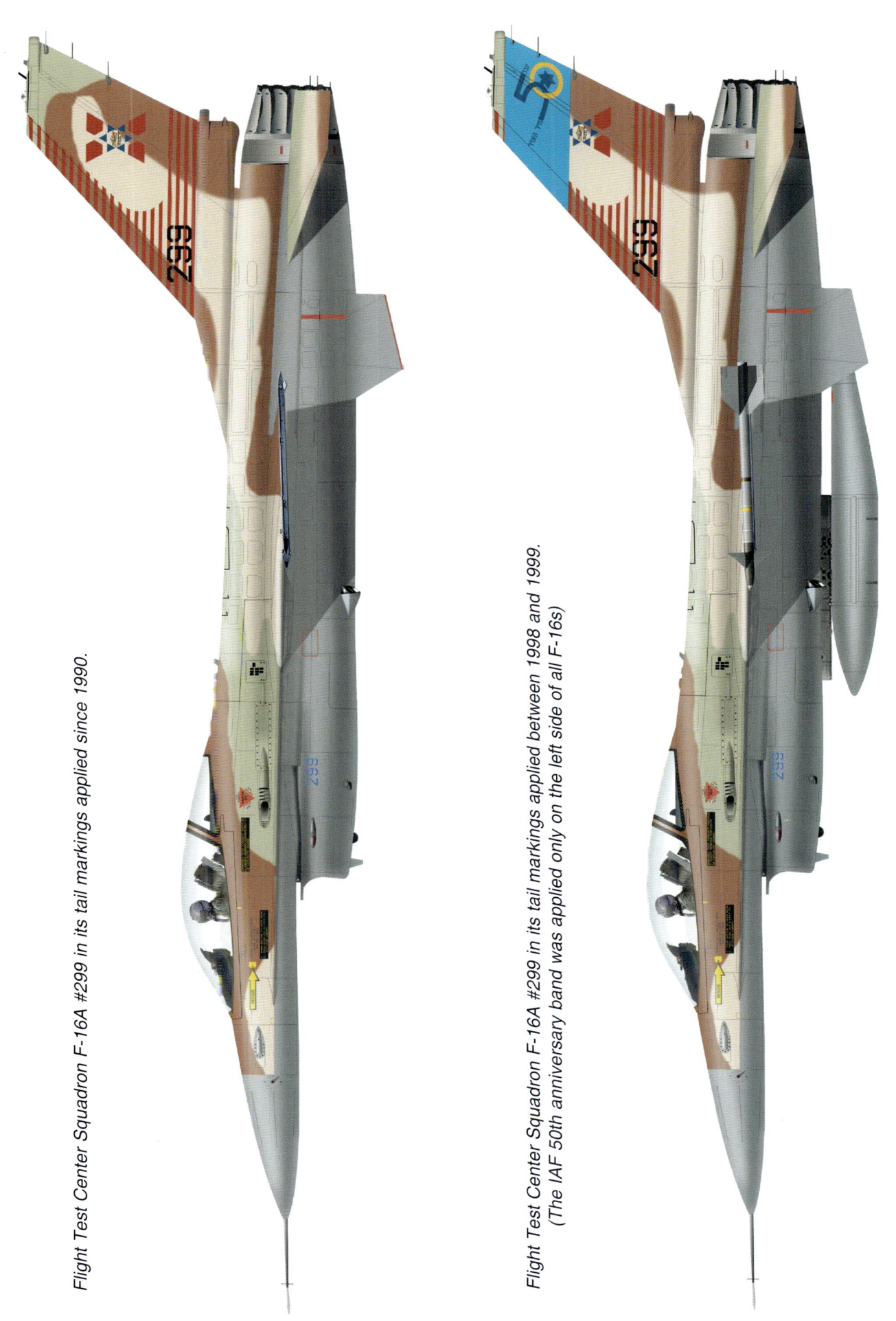

Flight Test Center Squadron F-16A #299 in its tail markings applied since 1990.

Flight Test Center Squadron F-16A #299 in its tail markings applied between 1998 and 1999. (The IAF 50th anniversary band was applied only on the left side of all F-16s)

First Jet Squadron Kill Table

DATE	PILOT	TAIL Nr.	WEAPON	KILL
28.4.1981	Rafi Berkovich	112	Canon	MI-8
28.4.1981	Dubi Yoffe	126	AIM-9	MI-8
21.4.1982	Hagai Katz	284	AIM-9	MiG 23
21.4.1982	Zeev Raz	107	AIM-9	MiG 23
9.6.1982	Eliezer Shkedi	107	AIM-9	MiG 23
9.6.1982	Eliezer Shkedi	107	AIM-9	1/2 MiG 23
9.6.1982	Eytan Stibbe	129	AIM-9	1/2 MiG 23
10.6.1982	Rafi Berkovich	116	AIM-9	MiG 23
10.6.1982	Rafi Berkovich	116	AIM-9	MiG 23
10.6.1982	Rafi Berkovich	116	Canon	MiG 21
10.6.1982	Itzhak (Sasha) Levin	111	AIM-9	MiG 21
10.6.1982	Itzhak (Sasha) Levin	111	AIM-9	Gazel
10.6.1982	Shlomo Zaytman	124	AIM-9	MiG 23
10.6.1982	Ami Lustig	138	Canon	MiG 23
10.6.1982	Hagai Katz	118	AIM-9	MiG 23
11/6/1982	Rani Falk	258	Canon	MiG 21
11/6/1982	Rani Falk	258	AIM-9	Su-22
11/6/1982	Amos Bar	252	Canon	Su-22
11/6/1982	Dan Oshrat	254	AIM-9	MiG 21
11/6/1982	Dan Oshrat	254	No weapon used	Su-22
11/6/1982	Eytan Stibbe	107	AIM-9	Su-22
11/6/1982	Eytan Stibbe	107	AIM-9	MiG 23
11/6/1982	Eytan Stibbe	107	AIM-9	Su-22
11/6/1982	Eytan Stibbe	107	AIM-9P3	Gazel
13.6.1985	Itzik Gat	112	Canon	DR-3

Knights of the North Squadron Kill Table

DATE	PILOT	TAIL Nr.	WEAPON	KILL
14.7.1981	Amir Nahumi	219	AIM-9	MiG 21
25.5.1982	Amos Mohar	240	AIM-9	MiG 21
25.5.1982	Amos Mohar	240	AIM-9	MiG 21
8.6.1982	Shlomo Sas	225	AIM-9	MiG 23
8.6.1982	Dubi Ofer	242	AIM-9	MiG 23
8.6.1982	Avishai Kna'an	250	AIM-9	MiG 23
9.6.1982	Relik Shapir	223	AIM-9	MiG 21
9.6.1982	Ofer Einav	237	AIM-9	MiG 21
9.6.1982	Amir Nahumi	220	AIM-9	MiG 21
9.6.1982	Avi Lavi	255	AIM-9	MiG 21
9.6.1982	Roee Tamir	250	AIM-9	MiG 21
9.6.1982	Avishai Kna'an	243	AIM-9	MiG 21
9.6.1982	Relik Shapir	232	AIM-9	MiG 21
10.6.1982	Amir Nahumi	234	AIM-9	MiG 23
10.6.1982	Amir Nahumi	234	Maneuver	MiG 23
10.6.1982	Amir Nahumi	234	AIM-9	MiG 21
11.6.1982	Relik Shapir	225	AIM-9	Su-22
11.6.1982	Relik Shapir	225	AIM-9P3	Su-22
11.6.1982	Yehuda Bavli	240	AIM-9	Su-22
11.6.1982	Amir Nahumi	237	AIM-9	MiG 21
11.6.1982	Amir Nahumi	237	AIM-9	MiG 21
11.6.1982	Roee Tamir	246	AIM-9	MiG 21

The Negev Squadron Kill Table

DATE	PILOT	TAIL Nr.	WEAPON	KILL
9.6.1982	Arnon Sharabi	287	AIM-9	1/2 MiG 23
9.6.1982	Moshe Rozenfeld	272	AIM-9	1/2 MiG 23
9.6.1982	Moshe Rozenfeld	272	AIM-9	MiG 21
9.6.1982	Nimrod Gur	267	AIM-9	MiG 23
9.6.1982	Uri Gil	290	AIM-9	MiG 23
9.6.1982	Uri Gil	290	Canon	MiG 21

Netz kill board, as presented at the closing ceremony of the Netz era in the Israeli Air Force. Nevatim AB June 4, 2015.

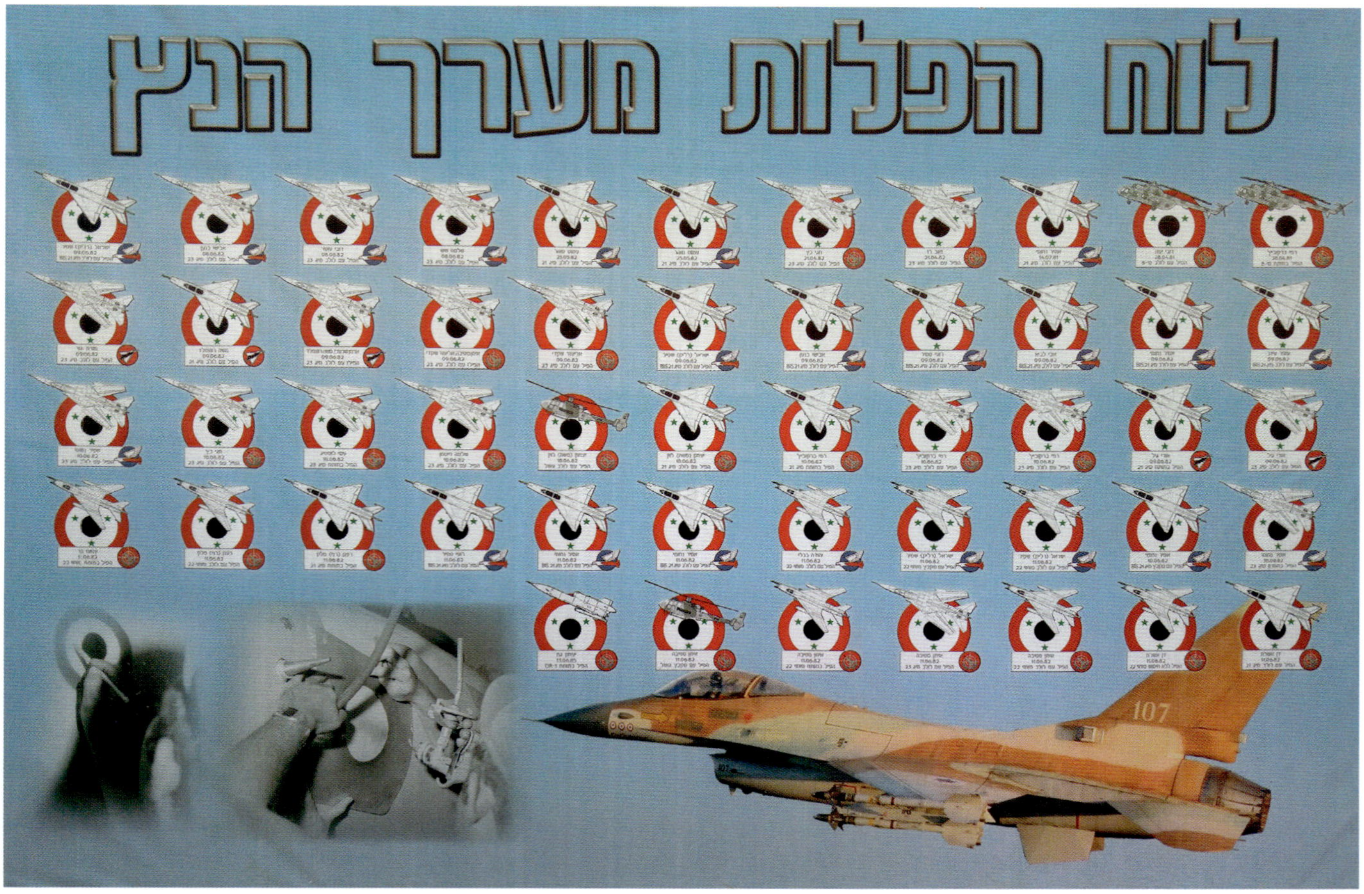

Closing ceremony of the Netz operational era in the Israeli Air Force. Nevatim AB June 4, 2015.

(Photos: Udi Burg & Adi Matos)

IAF Commander Major General Amir Eshel with Pini Elmakies, one of the 253 Squadron veterans. ▼

Photos on this page: Roy Amar

◄ *IAF Commander Major General Amir Eshel with 253 Squadron veterans.*

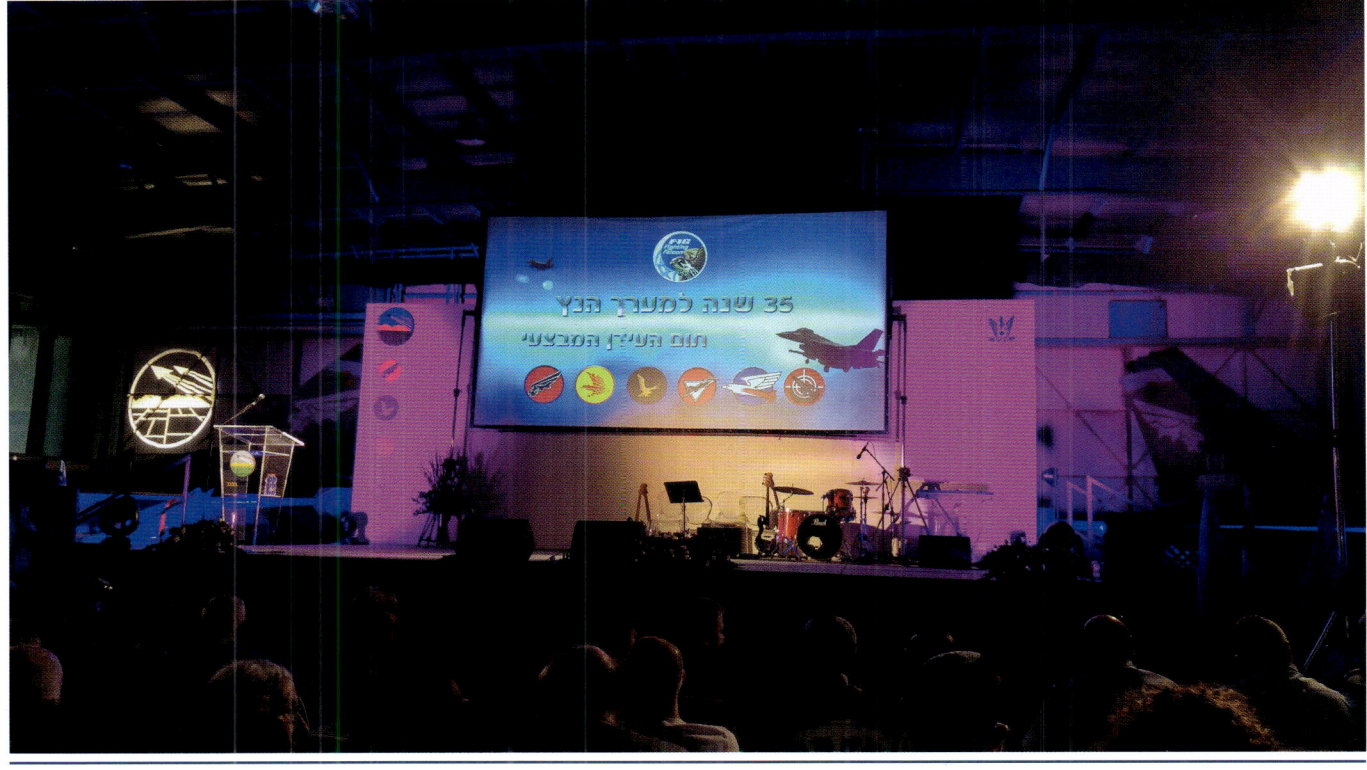

From the cockpit Photos: Major Ofer

Photo: Major Ofer

Other RN Publishing Books By the Author:
Israeli Eagles F-15A/B/C/D/I
ISBN: 978-88-95011-18-9

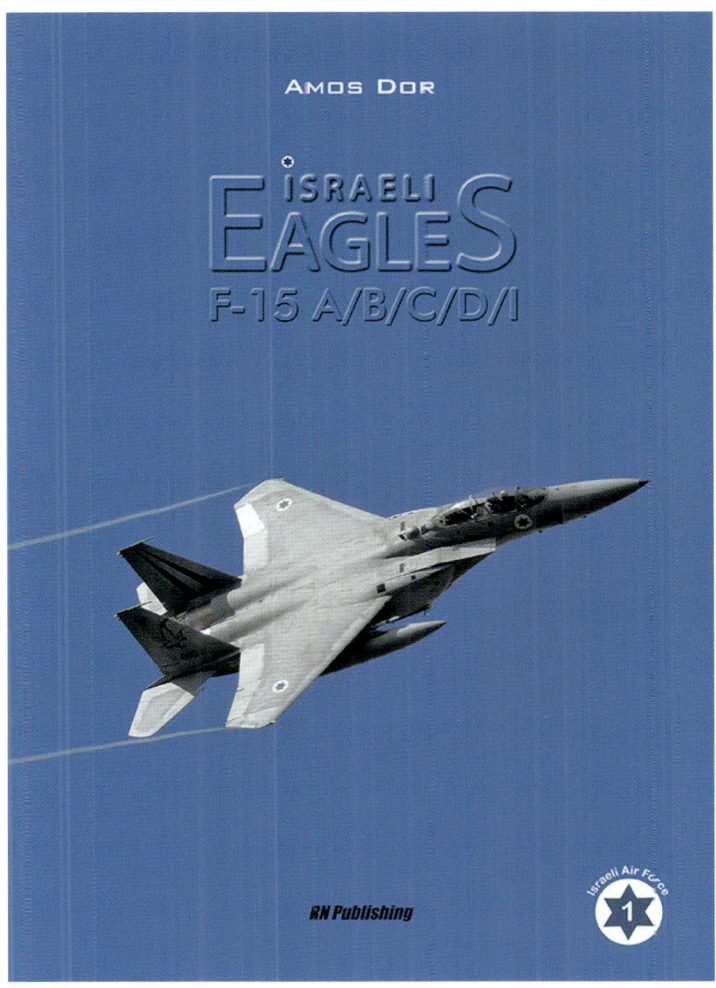

This first title open a new series dedicated to the most important aircraft operated by the Israel Air Force. The series starts with an icon which is also a myth in the aviation world: The F-15 eagle!

The book, which collect also unpublished photos, and information, is divided into several chapters, dedicated to the introduction in service of the aircraft, its use in peace and wartime operations with the first line squadrons, and its participation to several international air exercises. Besides several colour profiles, the book offers more than 500 photos, many of great interest, rare and unpublished, coming from private archives, that show all the aircraft operated, serial number by serial number. The book has been prepared by Amos Dor, a well known Israeli author, and comes after years of researches.

240 pages, text in English, more than 500 colour and b/w photos, 12 colour profiles.